ENZYME-THERAPY

ENZYME = THERAPY

By
MAX WOLF, M.D.,

And
KARL RANSBERGER, Ph.D.

Biological Research Institute, New York

VANTAGE PRESS

New York Washington Hollywood

FIRST EDITION

All rights reserved including the right of reproduction in whole or in part in any form

Copyright © 1972, by Max Wolf and Karl Ransberger

Published by Vantage Press, Inc.
516 West 34th Street, New York, New York 10001

Manufactured in the United States of America

Standard Book No.

Standard Book No. 533-00393-8

CONTENTS

	Page
Introduction	vii
Preface by Primarius Dr. Johannes Kretż, President of Austrian Cancer Society	xi
History and General Enzymology	13
Resorption of Enzymes	26
Biology of Inflammation	34
General Fibrinolysis and Thrombolysis	47
Experimental Thrombolysis	61
Thrombolytic Therapy with Streptokinase	69
Thrombolytic Therapy with Plasmin	78
Thrombolytic Therapy with Urokinase	80
Clinical Results of Enzymetherapy in Inflammations and Thromboses	84
Therapeutic Uses of Proteolytic Enzyme Mixture in Different Inflammations	93
Substitution Therapy by Digestive Enzymes	101
Aging and Enzymes	104
Virus-diseases and Enzyme Therapy	119
Enzyme Therapy of Cancer	135
Cancer Therapy with Specific Acting Enzymes	147
Pathology and Biochemistry of Metastasis Formation	150
Clinical Results of the Enzyme Therapy of Malignant Tumors—Unexpected Difficulties in Clinical Trials	167
Experiences and Opinions About Enzyme Therapy in Malignant Tumors	170
In Conclusion	221
Photographs	223
Index	230

INTRODUCTION

A very great number of articles have been published in the international literature about biochemistry, pharmacology and therapeutic use of enzymes. For the busy practitioner, the study of such publications to be found in different journals and different languages would be too tedious and often impossible. Important new investigations appear in special journals all over the world. They are for many not accessible, and we do not know of any up-to-date collection of established facts or textbooks giving the latest research on enzyme therapy.

A practical application of enzymes in the treatment of many different diseases or abnormalities depends on many items, not only on the knowledge of the late advances concerning biochemical processes, but also on experience in its application, the exact diagnosis, on continued monitoring and on interpretation of the results obtained. This may finally result in a picture of the possibilities as well as the limitations of enzyme therapy. Such critical attempts form the basis of proper use of individual drugs in general or of entire classes of medical products, like enzymes.

The monographs of *Innerfield* and of *Martin* about enzyme therapy are outdated to a great extent nowadays through the development of the latest research, and no new revised editions are planned. Our experiences gathered during the treatments with enzymes of several thousand patients as well as general research work on enzymes during the last twenty years, with the accumulated knowledge of the late world literature on it, induced us to write the present monograph. For instance, only during the recent years a close relation could be established between enzymes and inflammation, fibrinolysis, thrombolysis, cancer therapy and metastasis. Summaries about such connections will be given for the first time in this book.

Modern research of enzyme biochemistry and therapy occupies every year more and more space in the interest and armamentarium of the practising physician; we feel convinced that therapy with enzymes will gradually replace the present polypragmasy of the many symptomatic medicines derived from organic and inorganic chemistry, which flood the present drug market. To this end the present book may contribute material which may make today's physician familiar with the foundation, the present state of this science and the future possibilities of this so promising, relatively new branch of therapy. Here only several individual aspects of enzymology will be dealt with in detail. Of such an intricate science even a small branch can hardly be covered thoroughly in this book, much less the total science of practical enzymology.

This book endeavors to give the physician up-to-date information in his clinical work and the present state of research in some enzyme biochemistry. For one not familiar with enzymology the first chapter gives general facts about questions and problems in related biochemistry, about the types, formation and resorption of enzymes as well as analytical methods.

The first reports about application of proteolytic enzymes as biological anti-inflammatory agents were published in the United States. This therapy exceeds all other known pharmaceuticals. Many normal and pathological processes in the organism are induced and mediated by enzymes. Every enzyme has a specific action, for instance it splits or hydrolyses into definite substances. The continuous but labile equilibrium between enzyme and substrate during many important biological reaction and pathological deviations will be stressed on in this book, also the interaction between anabolic and catabolic processes.

Enzymes and most of their substrates are as a rule proteins of complicated chemical constitution, but some catabolize also many other organic compounds. This book deals mainly with the most important enzymes, the proteinases or proteases. Without going into complicated chemical formulas, we think it advisable to mention a few important biochemical and physical data determining the character and activity of the substances. Today we are using very purified, specific and well defined substances. Their tests and therapeutic application gave us numerous new insights and experiences.

On this occasion the authors wish to express their thanks to Dr. Kurt Maehder, Dr. Elizabeth Heuer and Dr. Otto Weigelt who helped on some items of the book.

For communications which may contribute to the practical use of protease therapy we should be grateful any time, also for suggestions of improvements or criticism of the other contents of this book.

New York (USA) and Munich (Germany) 1972

Max Wolf
Karl Ransberger

PREFACE

by *Primarius Dr. Johannes Kretz*

President of Austrian Cancer Society

The present monograph about proteolytic enzymes represents the first comprehensive work on the therapy with proteolytic enzymes.

In the introductory chapters also the reader less familiar with this material is informed in clear and concise form about general problems of the biochemistry of enzymes. The presence, formation and conditions of absorption are described and the different analytical tests discussed.

From the application of proteolytic enzymes in inflammatory processes which were first communicated by American authors, the therapeutic possibilities of this therapy spread far beyond the substitution therapy used so far.

The enzymes direct the normal as well as the pathological processes in the organism and influence to a great extent their course. In particular the modus operandi of the enzymes is discussed in separate chapters, during inflammatory processes, during fibrinolysis, at thrombolysis and during metastasis of malignant tumors.

In the section about the enzyme therapy of cancer the authors appreciate the discovery of the cytolytic properties of the normal serum against cancer cells by *Ernst Freund*. The proteolytic substance isolated by *Freund* from the serum, which he called "Normalsubstanz", has been used for parenteral therapy with partly good success on patients with advanced cancers. During the search of enzymes extracted from plants and animals which would be identical with the *Normalsubstanz* of *Freund*, it could later on be proven that the normal substance of *Freund* was indeed a proteolytic enzyme with cancercell-dissolving property.

Further chapters deal with the mutual relations between enzymes and their substrates during the biochemically important reactions of the normal organism, as well as during pathological deviations. Without going too much into details of the complicated processes of the enzymes and their substrates, the actions of the proteolytic enzymes are demonstrated.

Particularly valuable are the discussions about the biochemical effects of highly purified enzyme substances whose therapeutic applications have led already to numerous new additions of our knowledge.

HISTORY AND GENERAL ENZYMOLOGY

The harmonious interplay of all vital processes and their undisturbed functions is coordinated in the organism by the central, peripheral and the vegetative nervous system and is kept up by chemical and physical reactions, mediated through the circulation. In these processes the body enzymes take a prominent place, next to hormones and other active substances. In a variety of reactions they direct, accelerate, modify or retard all body functions. They are high-molecular, specific albuminous bodies found in every cell. They catalyze or hasten biochemical reactions which without them would be immeasurably slow. Enzymes are specific organic catalyzers.

Thus it can be easily seen that absence or insufficient amounts of one or several enzymes would necessarily lead to disturbance or functional failure of the cells or tissues involved. Abnormal synthesis or absence of an enzyme complex becomes manifest in the form of certain disease pictures (Enzymopathies).

Enzymatic processes have been known in antiquity. The manufacture of wine, beer, cheese, vinegar and other products of the daily life has been taking place with the help of enzymatic reactions. The Greeks ascribed the invention of wine to the god Bacchus. Over 20 centuries that mysterious transition called fermentation failed to become the object of research; the Arabic alchemists gave those unknown powers the name *Aliksir*. The Greek alchemist *Zozeen* who lived at the end of the third century in Egypt called them *Xerion*. In 1713 *Réaumur* examined the digestion of meat by the stomach juice of the buzzard in vitro and in vivo. These tests were resumed again in 1783 by *Spallanzani* who tested the stomach juice of crows. Naturally, about the mode of action of the digestive enzymes of the stomach no exact information could be gathered, though many scientists were later of the opinion that "certain" chemical decomposing reactions apparently are taking place by unknown substances in concentrations which could not be determined.

It was a giant step forward when *L. Pasteur* in the 19th century was able to prove that microorganisms, e.g. yeast, contain ferments or enzymes. Later on *Buchner,* among others, found that the enzymes are not bound to the living cells, but could be separated from tissues and cells by means of proper chemical or physical methods. The extracts have the identical specific enzyme action as the yeast itself, but are free from inert substances.

Previously several scientists described some enzymes and wrote the first rather vague notes about their activities. *Schwann* 1836: Pepsin of the stomach juice. *Liebig* and *Wöhler* 1837: Emulsin of bitter almonds, *Bussy* 1839: Myrosin in mustard, *Kühne* 1848: Trypsin of the pancreas juice, Cl. *Bernard* 1849: Lipases.

Of deciding influence for the chemical studies of enzymatic reactions were the investigations of the utilization of sugar in the living cells and its fermentation. The laboratories of the large breweries did valuable work in this respect. The fundamental chemical reactions during fermentation are almost the same as the glucose metabolism in the animal and plant cells.

These investigations paved the road to the production of pure enzymes. In 1837 *Berzelius,* who was the first scientist to study the chemistry of enzymes, clarified the fact that enzymes are catalysators of living cells. It took almost 100 years thereafter until *Sumner* in 1926 was able to produce the enzyme urease in pure and crystalline form. The work of *Sumner* was received very skeptically by the profession, until a few years later *Northrop* and *Kunitz* reported about the crystallization of trypsin, chymotrypsin and pepsin. This finally led to intense chemical studies of enzymatic reactions. A very large part of present publications about physiology and pathology of cells deals with enzymatic actions and reactions, for it is known that almost all physiological reactions, like muscle contraction, nerve conduction, urine excretion etc. are entirely conditioned by enzyme activities.

Up to now, several hundred different enzymes are known, about 75 of them can be had in crystalline form. All enzymes are protein bodies. By means of crystalline enzymes it became possible to get insight into the structure of protein bodies in general. Through the most modern methods, like gas chromatography, ultracentrifugation, X-ray diffraction etc. the amino acid or peptide-binding sequences of protein building blocks, molecular weight, as well as the intracellular activity of enzymes in vivo and vitro were investigated and defined. Enzymes became the standard pattern for structure-analytical and physical investigations of proteins in general.

The time-related development of enzymatic splitting processes showed that the same enzymes in proper concentration and environment are also able to synthetise or polymerize. With a great many synthetic substrates the substrate specificity was determined which was also important for classification and identification of the individual enzymes. The development of enzymology during these years is an example of research activities on a world-wide scale. They continue now for many years among individual research groups, irrespective of political circumstances, and to a greater extent than with other research problems in natural science. It is an inspiring triumph of the human mind and human collaboration. All enzymes which are splitting or catabolic are under different conditions anabolic or polymerising and build up larger molecules, as mentioned before.

It stands to reason that such a young science, particularly a product of different disciplines, has at first to create its own language, a "nomenclature", in order to make an international understanding possible. It makes further research much easier if uniform names (or symbols) are given the important enzymes, and the methods of their determination has been standardised and internationally agreed upon.

How remarkably the progress of modern enzymology has advanced can be seen in the following example. The very complicated and complex process of muscle contraction can be divided into a number of different enzymatically catalyzed part reactions. Most of these can now be separately analyzed and induced. This technique is so far advanced that the effect of each successive enzyme reaction in vitro is identical with the physiology of the live muscle fiber.

The speed of each simple reaction can be increased by enzymes, or the reaction does not take place at all without catalyzer. For instance an exchange of carbon dioxide between blood and tissues or blood and lungs without the enzyme carboanhydrase would be entirely impossible because this reaction without catalyzers would take place immeasurably slowly, the exchange of the gases would promptly stop.

Enzymes are high-molecular, thermolabile proteins which make almost all chemical reactions in the body (metabolism) possible. Most of them are built up of two parts, the apoenzyme of protein nature and the co-enzyme or the prosthetic group, a usually small molecule, like a vitamin. Both together give the active enzyme, called holo-enzyme; each separate part is inactive as a rule.

All enzymes are characterized by certain substrate specificity which is mostly determined by the apoenzyme. Different from the catalysts in inorganic chemistry, enzymes not only accelerate reactions but are able to bring them about which without them would not take place at all. Another difference from inorganic catalysts is the fact that during enzyme activity a certain amount of it is lost, while inorganic catalysts can be totally recovered or regenerated after the reaction.

Classification of Enzymes:

Different classifications are possible, depending on the scientific viewpoint, on chemical, bio-chemical or other data. Here we restrict the terminology to few classifications. Several longer known enzymes have trivial names. like trypsin, pepsin. Others were named after their sources, like pancreatin, papain. Most enzymes are characterized by the endsyllable "ase", the main part of the name usually refers to the substrate: proteases split proteins, nucleases split nucleic acids. The exact name of some enzymes also gives the type of action besides the substrate, e.g. lactose-dehydrogenase, cytochrome-oxydase.

Nomenclature:

Enzymes are usually named in terms of the reactions which are catalyzed. The suffix -ase is added to the name of the substrate. Enzymes may also be classified by groups or classes, as the proteinases, lipases, oxidases etc. According to the "Report of the Commission on Enzymes of the International Union of Biochemistry" enzymes can be divided into 6 main groups as follows:

1. Oxidoreductases
2. Transferases

3. Hydrolases
4. Lyases
5. Isomerases
6. Ligases (= Synthetases)

The oxidoreductases concerned with oxidation-reduction processes play an important role in the metabolism.

The transfering enzymes (transferases) catalyze the transfer of a group from one substrate to another. Best known are the transamidases which are of real value in medical diagnostics.

Hydrolytic enzymes (or hydrolases) act by catalysing the introduction of water at a specific bond of the substrate. Many groups of them are known: esterases, phosphatases, nucleases etc.

Lyases are enzymes which remove groups from their substrates, but not by hydrolysis, or conversely add groups to double bonds.

Isomerases catalyse the interconversion of aldose to ketose sugars.

Ligases are enzymes which catalyse the joining together of two molecules, coupled with the breakdown of a pyrophosphate bond or a similar triphosphate.

Enzyme Actions:

The action of almost all enzymes is brought about by entering into a temporary chemical union with its substrate, thereby changing it, and afterwards separating themselves from it. After this reaction the enzymes stand ready again for an analogous repetition of the process. Expressed in form of a formula, the hydrolase effect would be as follows:

$$AB + E \text{ (enzyme)} + HOH \longrightarrow ABE + HOH \longrightarrow AH + BOH + E$$

$$E + S \underset{K_2}{\overset{K_1}{\rightleftarrows}} (ES) \overset{K_3}{\longrightarrow} E + P \text{ (Michaelis)}$$

The basic equations for the reaction of an enzyme and its substrate were developed by *Michaelis* and *Menten*. In this reaction mechanism, a substrate S combines with the enzyme E to form an intermediate complex (ES), which subsequently breaks down into the split product P and liberates the enzyme E. The enzyme is a true catalyst, since it is not consumed in the reaction.

The equilibrium constant for the formation of the complex (ES) is called the Michaelis constant (Mk), which is defined as $\frac{(K_2 + K_3)}{K_1}$. It gives the speed of the reaction and is dependent on substrate- and enzyme-concentration. Like most biochemical reactions, enzymatic processes are a matter of equilibrium. The chief factors which determine the initial velocity of an enzymatic reaction are: enzyme concentration, substrate concentration, pH range, temperature, activators, inhibitors and ionic strength.

Methods of Assay:

Most assays are based on determination of the enzyme activity by letting the enzyme solution act under standardized conditions on its substrate. The reaction products can be estimated by manometric-, spectrophotometric-, potentiometric-, polarographic-, fluorescence or radiochemical methods.

The enzyme quantity is given in units which are internationally defined for some enzymes. Unfortunately many different units have been proposed for the same enzyme by different methods. Many authors use and publish test methods according to their own suggested standards or methods; this causes confusion. An exact comparison or transfiguring of the different units is impossible.

In order to avoid confusion, 1959 the Enzyme Commission of the International Union of Biochemistry and the Section of Clinical Chemistry of the International Union of Pure and Aplied Chemistry accepted a new "international unit". This unit is defined as the amount of enzyme which converts one (micro) mole of substrate per min. at 25° under optimal substrate concentration, pH and ionic strength of buffer systems. The specific activity of the enzymes is defined as the activity of 1 mg of the enzyme. Enzymes can also be determined in tissue sections (histochemistry).

Enzyme determinations are of great importance in the diagnosis of diseases and in monitoring progress of disease. Enzymes attack mostly only one component of racematic substances, e.g. the one existing in nature. This specificity is important in stereochemistry because thus racemates can be separated into optical antipodes.

Enzyme Activators and Inhibitors

Chemical substances or forces which increase enzyme activity independent of ions, temperature, pressure or substrate concentration, are called enzyme activators, those which reduce it are called inhibitors or antiactivators. The activator can be effective by two different means: either it raises the number of "active groups" of the enzyme molecule (genuine activators) or it checks the inhibitors (antiinhibitor).

The activators play an important part in enzyme therapy. Their effect is brought about by the fact that they join an active group of some part of the apoenzyme or co-enzyme or both. This union may be reversible or irreversible.

One type of inhibitors acts by so-called pseudosubstrates. These are substrates which very much resemble the normal substrate. The enzyme molecules enter in such a case into a temporary or a lasting union with the pseudosubstrate and thus the enzyme cannot attack any more the normal substrate itself. About individual enzyme inhibitors will be reported later on, they are very important for all enzyme physiology.

Isoenzymes:

Isoenzymes (or isozymes) are defined as multiple forms of enzymes occurring in the same organism and having similar or identical catalytic activities, differing only in biophysical constants. They are usually demonstrated by the techniques of zone electrophoresis followed by in situ histochemically staining for the specific activity.

Immobilized enzymes:

Enzymes could be prepared in an immobilized (insolubilized) form without loss of activity. These enzymes can be stored on elevated temperature for months or years without any loss of activity.

Two major techniques can be used to immobilize an enzyme: 1) the chemical modification of the enzyme molecule and 2) the physical entrapment of the enzyme in an inert matrix, such as starch or polyacrylamide gel. The immobilized enzymes will likely bring a new future to biochemistry and enzyme analyses.

Applications for immobilized trypsin or chymotrypsin are: The possibility of the isolation of specific trypsin- or chymotrypsin inhibitors, using columns containing the insolubilized preparations, the separation of specific antibodies to enzymes in immunochemistry.

The table shows an insolubilized trypsin, imbedded in a copolymer of maleic anhydride and ethylene, cross-linked with hexamethylene-diamine.

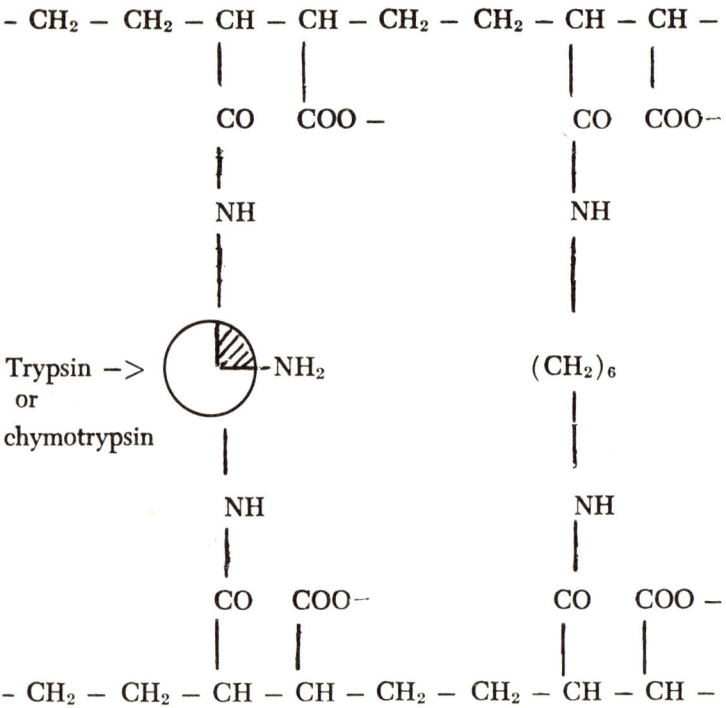

Biochemistry of Enzymes:
Since crystalline enzymes, as highly purified substances, have been available, enzyme research developed very rapidly. In the limited framework of this monograph, we confine our main interest to the proteolytic enzymes and lipases. For therapeutic application the hydrolases are the most interesting. Protein-splitting enzymes are mainly the proteases, e.g. trypsin, chymotrypsin, plasmin (fibrinolysin), pepsin, papain, bromelin, ficin, fungal proteases and a number of others, extracted from tissues.

To the estersplitting enzymes belong mainly the lipases, specific and non-specific esterases, phosphatases and nucleases. Of the glycosid splitting enzymes the most important representatives are the amylases. An interesting new enzyme is the lysozyme (3.2.1.17) (synonym:muramidase). Found in egg white, tears etc., it enhances the activity of antibacterial compounds, e.g. antibiotics).

Trypsin and Chymotrypsin (C E. 3.4.4.4., 3.4.4.5. and 3.4.4.6.)

Trypsin and chymotrypsin A and B are endopeptidases and split central peptide links from proteins and substituted peptides, mostly at certain points of attacks.

Chymotrypsin is formed in the duodenal juice. From the beef pancreas two inactive proenzymes can be isolated: alpha- and beta-chymotrypsinogen, which are activated by trypsin to the active enzyme. The molecular weight of chymotrypsinogen is 24.000, of chymotrypsin 22.000.

All these groups split proteins into peptides, but also esters and amides of aromatic amino acids *(Neurath, Dixon)*. Chymotrypsin splits denatured proteins, preferably the peptide links of aromatic amino acids (carboxyl ends); also at a slower rate leucyl − methionyl − norvaline − asparaginyl and glutaminyl compounds are hydrolyzed.

Trypsin is, like chymotrypsin, a proteolytic enzyme, with trypsinogen as proenzyme. It is also formed in the pancreas. It has a molecular weight of 24.000. The proenzyme is changed into the active enzyme in the body by trypsin or enterokinase, which is formed in the mucous membrane of the intestinal tract. It splits a hexapeptide off the trypsinogen.

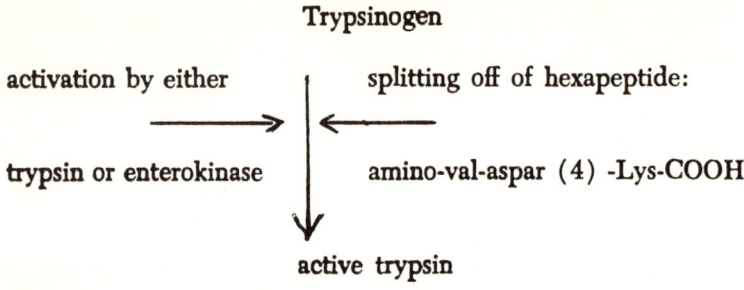

Neurath, Turba

During this activation also an inactive *"inert protein"* is formed which is regarded as polymer of the primary enzyme (*Bier*). Calcium ions check the formation of the inert protein (*McDonald, Gorini*). By electrophoresis at least 4 fractions of the active trypsin can be separated which are distinguished by the symbols F_1 to F_4.

Trypsin splits denatured proteins specifically at the carboxyl group of arginine.

For the determination of the activity in vitro are specially suitable: N-alpha-benzoyl-phenylalanine-beta-naphthyl ester (*Ravin*) and L-tyrosine-acetyl-ester, or the corresponding N-acetyl ester (*Schwert*).

During the hydrolysis of peptides, amides or esters of these amino-acids the speed of reactions increases in the order of peptide—amide-ester.

Pepsin:
In the fundus glands of the stomach 1 g. pepsinogen is formed daily whose activation to pepsin takes place by the hydrochloric acid of the stomach. The molecular weight of pepsinogen is 42.000, of pepsin 35.000. The pH optimum is 1.5 to 2.

The main part of pepsin is secreted into the stomach and there takes an active part in the process of digestion. Other amounts of pepsinogen get into the circulation and are eliminated through the kidneys.

Pepsin splits almost all proteins with the exception of some protamines. Also synthetic peptides are hydrolyzed, if on both sides of the point of splitting a L-configuration exists. The specificity toward rest of amino acids is slight, however, aromatic rings are preferred.

Cathepsin: (C.E. for Cathepsin C 3.4.4.9,
for Cathepsin D 3.4.4.23)

Cathepsin is an endocellular protease and is formed in several tissues. It consists of at least four single enzymes with different molecular weights. The pH optimum is 5. The main function of cathepsin is the autolysis of cells at the "cell moulting". It is activated by glutathion, cystein, prussic acid or H_2S, it is inhibited by copper, iodo-acetic acid and phenylhydrazin. On account of the low substrate specificity almost all proteins and their split-products down to amino acids are hydrolyzed.

Papain and Papainases:

Papain is produced from the leaves and fruits of Carica papaya. It has a molecular weight of 20.500, its pH optimum is about 5.5. Glutathion, cystein and thiosulfate are activating, copper and peroxide of hydrogen inhibiting. The substrate specificity is slight because many proteins and synthetic substrates are split.

Lipases:

Lipases are extracted from pancreas and other tissues, also from fungi (aspergillus orycae). The pH optimum of the pancreas lipase is about 9. This lipase is activated by glutathion, cystein and potassium cyanide, and inhibited by quinine, copper, iron or aldehydes. The physiological part played by the lipases in the blood is not very clear; probably they are concerned with the lysis of fat deposits and utilization of the fats. Lipases split fats into fatty acids, especially the glycerin esters, in contrast to the esterases whose substrates are the short chain fatty acid esters.

Amylases:
Amylases are found in many organs (liver, pancreas) and in different fungi. They split starch and glycogen to maltose at a pH optimum of about 6.5-7.2.
The alpha- and beta-amylases are best investigated and have a molecular weight of about 45.000.

Use of Enzymes:
Enzymes found a broad area of application in physiological and technical chemistry. Of the great number of possibilities here only a few fields of application of proteolytic enzymes may be mentioned:
1) in chemistry: for the analysis of the chemical constitution of unknown proteins and proteohormones.
2) in pharmacy: Antiinflammatory agents, fibrinolytic, carcinostatic and cytotoxic compounds. As substitution therapy in digestive disturbances, in diseases caused by worms.
3) in industry: Active constituents of substances of laundry use (in contrast to the animal proteases, the plant proteases used for laundry work are temperature-stable and can be heated to 90 C° without any appreciable loss of activity). As means in the tanning and food industry (e.g. for tenderizing meat).

From animal experiments and clinical experiences can be concluded that proteolytic enzymes, even in large doses, are practically nontoxic when given by oral, rectal or i.m. routes.

Comprehensive informations about chemistry and assay methods of enzymes are found in the following textbooks:

Umbreit. W. W., R .H. Burris, J. F. Stauffer,
Manometric Techniques, Burgess Publ. Co., Minneapolis, 1945

Boyer, P. D., H. Hardy, K. Myrbäck,
The Enzymes, Vol. 1-6, N.Y., 1962

Kolowick, S. P.
Methods in Enzymology, Vol. 1 and 2, Academic Press, N.Y., 1955

Dixon, M., C. E. Webb,
Enzymes. Longmans, Green, 1958

Abderhalden, R.,
Klinische Enzymologie, Thieme, Stuttgart, 1958

Hoffman-Ostenhof, O.,
Enzymologie, Wien, 1954

Northrop, J. H., M. Kunitz, R. M. Herriott,
Crystalline Enzymes, Columbia Univ. Press, N.Y., 1948

Ambrose, A. M.,
Naturally Occurring Antienzymes,
Natl. Acad. Science Nav. Res. Council Nr. 1354, 1966

Bergmeyer, H. U.,
Methoden der enzymatischen Analyse, Verlag Chemie, Heidelberg, 1968

RESORPTION OF ENZYMES

Enzymes have aroused considerable therapeutic interest in inflammatory conditions since the investigations of *Innerfield* and others. In this connection the question of absorption after parenteral, oral and local application became important and found its expression in numerous publications.

The resorption of enzymes by parenteral route is certainly an established fact and does not need further discussion.

Investigation about the resorption of proteolytic and other enzymes after oral or rectal application led to various and partly contradictory results. This is certainly not only caused by the different methods of investigation, but also to a great extent by the difficulties resulting from the determination methods of proteolytic activity in the serum with antiproteases. While on account of these facts one can hardly expect unequivocal results in regard to dosage and resorption, the investigations of the last years prove that by activation of the endogenous enzyme a distinct pharmacological effect or an antiinflammatory process is taking place. *Smyth* emphasizes that the clinical results after oral enzyme doses can be readily explained without assumption of a resorption of exogenous enzymes from the digestive tract. Late experiments allowed the conclusion that the enzymes, perhaps at the place of application (gut), induce the formation of secondary enzymes or catalyze activators or other biologically active substances which are capable of being resorbed because their molecular structure (smaller molecular weight) differs from the former. The resorption of fragments of the enzyme molecule is already sufficient for a pharmaceutical effect.

For several years it has been possible to investigate questions of resorption or metabolism by means of radio-active substances. Thus *Miller* and *Martin* used tagged trypsin and chymotrypsin marked by J-131. The oral doses of these substances caused a significant increase of proteolytic activity of the serum. *Miller* presumes that the marked compound becomes attached to the albuminous bodies of the serum and he concludes that it must be the marked enzyme itself, but a direct proof of the resorption is not established by it. The marking with J-131 is only slight and the only loosely bound iodine can very rapidly again be split off. The possibility also cannot be excluded that in vivo other albuminous bodies could be secondarily marked (transmarking, an analogous reaction to transamidation) and that the presence of trypsin may be only a deception.

The chemical marking of albuminous bodies with fluorescent dyes as used in the immunochemistry is another possibility to prove resorption. *Nairin* suggests to join enzymes directly to fluorescent dyes. *Kay* announced a similar method of binding the enzymes to cells by means of chloro-s-triazines (28). Bromelin forms with nitrophenol a stable compound which can be tested for after oral administration in the serum after 1½ hours to 69% and after 2 hours to 86% (*Seligman* (45). *Smyth* uses for his investigations bromelin which is changed into a strongly fluorescent protein-dye complex by means of 5-dimethyl-amino-naphthole-sulfochloride which can easily be tested for in the serum and in organs (50). 10 to 50 mg. of this complex have been given to rabbits. Also here was the principal question whether the testing of fluorescence of serum or tissues can determine the presence of the intact enzyme-dye-complex. Electrophoretic investigations showed without doubt that the unchanged active enzyme molecule joined to the dye could be established in the serum. This gives a proof of the oral absorption. At the end of the experiment the animals were killed and the fluorescence investigated. It could be found in liver, kidney and urine but not in lungs, heart, spleen or duodenum.

Smyth emphasizes at the end of his article that the question of proof of oral resorption of the enzyme molecule is primarily only of academic or scientific interest, its answer is fundamentally not important for the pharmacological effect. He says: "The proteolytic enzyme can just as well act by the formation of secondary enzymes or other biologically active compounds in the intestines which when absorbed from there would inhibit inflammation" (50).

Contradictory are the results reported by individual authors, but most agree that trypsin acts after intramuscular, buccal, rectal or oral administration by reducing inflammation and edema in experimental inflammations as well as in clinical cases [*Martin, Veermenko, Bare, Innerfield* etc. (2, 34, 35, 36)]. For the determination of trypsin in the serum many methods are suggested (*Nardi* (37), *Hummel* (23), *Seligman, Kallos* (27), *Babson, Roth, Ronwin* (42, 43) and *Hestrin* (22)). Several investigators could demonstrate tryptic activity in the normal human plasma (*Ronwin* (43)) or serum (*Kallos* and *Seligman* (26)).

Megel describes a sensitive method for the demonstration of trypsin in plasma which *Kabacoff* demonstrated already in principle for the determination of chymotrypsin. Already in the normal untreated rat-plasma two or more enzymes could be found which split BAEE. The one enzyme resembles trypsin, for it splits BAEE and is inhibited by SBI (140 gamma trypsin are inhibited ca. 97% by 90 gamma SBI), the other enzyme (or several ones) hydrolyzes BAEE, but is not inhibited by SBI. Plasmin and ribonuclease do not attack BAEE while chymotrypsin hydrolyzes it slightly. After oral administration of 500 mg./kg. trypsin in oil or in an emulgator (for protection) the use of the *Megel* method showed that the plasma-trypsin-activity is significantly increased compared to the controls. If an equal dose of enzyme is given in physiological salt solution, the plasma enzyme level is not influenced. These facts confirm and enlarge the results of *Martin*. He found that 20 mg/kg trypsin, introduced directly into the ilium can check an experimental inflammation. 200 mg/kg oral in an emulgator have the same effect, while even 500 mg/kg enzyme in salt solution do not show any influence.

Plasma-trypsin and total esterases 45 minutes after peroral application of trypsin in oil in rabbits (*Megel* (53)) give the following results:

trypsin mg/kg	plasma-trypsin gamma trypsin o, 1 mg	total esterase gamma o, 1 ml
control	1.51 + 0.09	3.77 + 0.09
125	1.63 + 0.21	3.80 + 0.06
250	1.60 + 0.21	3.82 + 0.11
500	2.65 + 6.25	4.54 + 0.16

The proteolytic enzymes are resorbed from the gastro-intestinal tract. However, since large amounts of the enzymes are inactivated there, large initial doses are required. A sufficient enzyme niveau was only demonstrable after continuous infusion of high quantities of enzymes (*Sherry*). If the pharmacological effect is not brought about by a primary resorption, it is to be presumed that the exogenous introduced enzymes stimulate the formation of other plasma enzymes of the body which in this case hydrolyze BAEE and are inhibited by SBI (*Megel*). Another possibility exists that the exogenous enzymes check or inhibit trypsin inhibitors present in the organism. According to *Megel* it is however more probable that the intact enzyme is resorbed directly through the mucous membrane.

Kabacoff studied the reaction on the level of plasma enzymes of rabbits after supplying them with 10 mg/kg chymotrypsin intramuscular, 20 mg/kg intestinal and 20 mg/kg per rectum (25). The normal amount of chymotrypsin activity in the serum (substrate ATEE) is 0.5 to 1.0 gamma/10 ml serum and is therefore extremely low. After all these kinds of application, the serum enzyme level was raised significantly. Since chymotrypsin splits ATEE and BPNA but does not attack TAME and BAEE, this established the proof that the splitting was produced by chymotrypsin and not by plasmin, thrombin, urokinase or any other esterase.

Abbreviations for substrates and inhibitors:
ATEE = N-acetyl-L-tyrosine ethylester
BPNE = Benzoyl-phenylalanine naphthyl-ester
TAME = Tosyl-arginine methylester
BAEE = Benzoyl-arginine ethylester
SBI = Soybean-trypsin-inhibitor

The relation of activities in vitro of ATEE to BPNE is 1,82 to 0,16, after rectal application of chymotrypsin it was 1,78 to 0,49. From the equality of this relation we may conclude that chymotrypsin is reabsorbed from the gastro-intestinal tract in enzymatically active form. Investigations on healthy test-persons showed that after rectal chymotrypsin application the enzyme is absorbed in active form through the gut (*Kabacoff* (24)).

These observations are supplemented by the investigations of *Avakian*. 27 patients received 80 mg chymotrypsin in enteric-coated capsules. After 2, 4 and 6 hours the proteolytic activity of the serum was tested. 4 hours after taking the enzymes the activity curve reached a maximum, thereafter it dropped rapidly (1).

Plasma enzyme level of rabbit after taking chymotrypsin (according to *Kabacoff*).

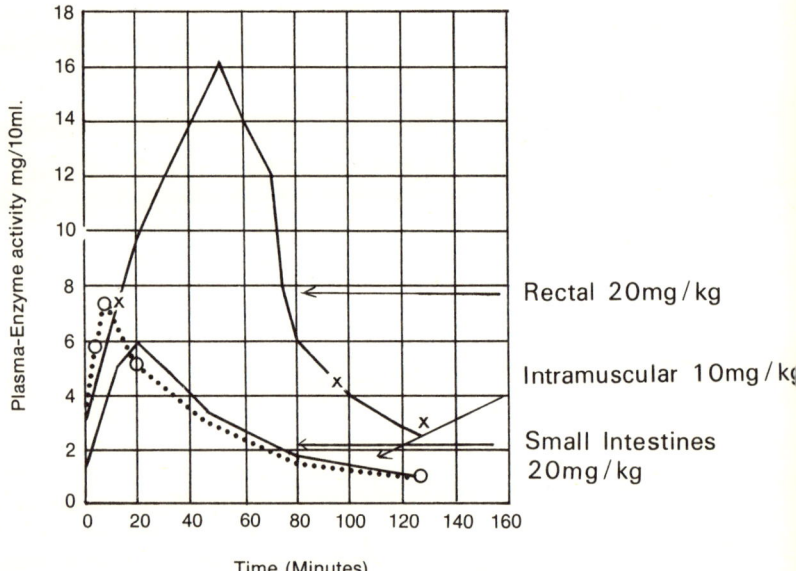

The absence of special outspoken biochemical changes or reactions in the body after oral administration of proteolytic enzymes caused a certain skepsis against this form of therapy which persisted in spite of the good clinical results. The pharmacological foundations of the parenteral, oral and rectal application of proteases are depolymerization and a gradual restitution to normal physiological condition of permeability (*Martin* (34,35,36)).

This broad principle of activity includes that the enzymes become effective either directly or indirectly. *Innerfield* and his group were the first to prove that the antithrombin activity in the plasma increases after oral administration of certain forms of proteases. The reaction is depending on dosage because small doses increase, very large doses lower the antithrombin level.

Smyth describes in detail the biochemical alterations which can be shown after orally taken bromelin (50). Rabbits, after receiving 5 mg/kg orally, showed 30 minutes after ingestion a prolongation of the prothrombin time up to 250%; similarly the level of plasmin in the serum was raised.

Plasmin concentration of the serum after oral intake of bromelin (rabbit):

Bromelin mg	Plasmin concentr. (Units/ml)		Time after intake (min)		
	0	30	60	150	270
100	4.0	8.0	10.4	17.7	10.6
25	5.6	10.7	15.5	16.8	10.0
15	4.4	7.6	9.2	6.4	5.6

A raised level of plasmin could accelerate the breaking down of fibrin. Quantitative tests proved that 1 mg bromelin activated as much plasminogen as 4000 U streptokinase.

Besides exogenous proteases like bromelin or trypsin, several endogenous proteases of the body proper became known during the last years by investigations of *Werle, Kraut* etc., e.g. kinines which are activated by exogenous enzymes. Bradykinine is a mediator of a general inflammatory reaction; *Innerfield* studied the influence of orally given proteases (papain, streptokinase etc.) upon bradykinine. The testing of the kinine activity on the uterus proved that bradykinine (and other kinines) are inhibited by these enzymes. This, by another way, is a proof for the oral resorption of enzymes.

The different methods of experimentally set inflammations (tests on rat-paw edema, croton oil injection, granuloma pouch, formalin edema etc.) respond in varying degrees to biological or chemical inflammation inhibitors. The most favorable influence, with the granuloma pouch test, is after parenteral administration of proteases which is agreed upon by most investigations (e.g. *Ungar, Sherry, Cohen, Naylor-Foote*). *Innerfield* was able to positively influence hepatitis produced by intradermally injected carbontetrachloride through oral intake of proteases.

In order to prove that the anti-inflammatory effect is caused only by the proteolytic activity, trypsin and chymotrypsin were inactivated by epsilon-amino-caproic acid. An enzyme thus inactivated did not have any influence on a model inflammation (*Innerfield* (54)).

Literature: Resorption of Enzymes

1. *AVAKIAN*, G., Clin. Pharmac. Therap., 5, 712 (1964)
2. *BARE*, W. W., Amer. J. Obst. Gynec., 87, 268 (1963)
3. *BARNETT*, C. A., Proc., 119, 866 (1965)
4. *BEILER*, J. M., et al., Proc., 89, 274 (1955)
5. *BELITSEV*, V. A., et al., Probl. Molec. Biol., USSR, p. 177 (1965)
6. *BOISSONAS*, R. A., et al., Helv. Chim. Acta, 43, 1349 (1960)
7. *CHAMBERS*, D. A., et al., Nature, 197, 1300 (1963)
8. *DARZYNKIWIECKZ*, Z., et al., Nature, 213, 1198 (1967)
9. *DIDESHEIM*, P., et al., Proc., 93, 10 (1956)
10. *DIETZ*, A. A., et al., Clin. Chim. Acta., Amsterdam, 13, 242 (1967)
11. *DINIZ*, A. R., et al., Biochem. Biophys. Res. Comm., 21, 488 (1965)
12. *DINIZ*, C. R., et al., Hypotensive Peptides, Springer, Berlin, 175 (1966)
13. *DYCK*, W., et al., Acta Gastroenterol. belg., 27, 590 (1964)
14. *ERDÖS*, E. G., et al., Hypotensive Peptides, Springer, Berlin, 235 (1966)
15. *FOLK*, J. E., et al., J. Biol. Chem., 240, 181 (1965)
16. *GILFOIL*, T. M., et al., Brit. J. Pharmacy, 27, 120 (1966)
17. *GREEN*, J. P., Arch. Intern. Pharmacodyn., 92, 1 (1952)
18. *GULLICK*, H. D., New England J. Med., 268, 851 (1963)
19. *HAUSTEIN*, K. O., et al., Arch. Intern. Pharmacodyn., 163, 393 (1966)
20. *HEIDE*, K., et al., Clin. Chim. Acta, Amsterdam, 11, 82 (1965)
21. *HENDLEY*, C. D., et al., Arch. Intern. Pharmacodyn., 106, 164 (1956)
22. *HESTRIN*, S., J. Biol. Chem., 180, 2496 (1949)

23. *HUMMEL*, B. C., et al., Canad. J. Biochem. Physiol., *37*, 1393 (1959)
24. *KABACOFF*, B. L., et al., Nature H, *199*, 815 (1963)
25. *KABACOFF*, B. L., et al., J. Pharm. Sci., *52*, 1300 (1963)
26. *KALLOS*, J., et al., Canad. J. Biochem., *42*, 235 (1964)
27. *KAMINURA*, M., Sapporo Igaku Zashi, *20*, 311 (1961)
28. *KAY*, G., et al., Nature, 216 (1967)
29. *KELLNER*, A., et al., J. Exp. Med., *99*, 387 (1954)
30. *KLINE*, D. L., Fed. Proc., *25*, 31 (1966)
31. *LAUWERS*, A., et al., Arch. Intern. Physiol. Biochim., *73*, 530 (1965)
32. *MACKAY*, A. V. P., et al., J. Clin. Pathol., *20*, 227 (1967)
33. *MANN*, R. D., J. Clin. Pathol., *20*, 223 (1967)
34. *MARTIN*, G. J., et al., Proc., *86*, 636 (1954)
35. *MARTIN*, G. J., et al., Amer. J. Pharm., *129*, 194 (1957)
36. *MARTIN*, G. J., et al., Exptl. Med. Surg., Suppl. *23*, 150 (1965)
37. *NARDI*, G. L., et al., J. Lab Clin. Med., *52*, 66 (1958)
38. *ÖGSTON*, D., et al., Thromb. Diathes. haemorrh., *16*, 32 (1966)
39. *PISTAN*, J., et al., J. Biol. Chem., *241*, 5090 (1966)
40. *RIESER*, P., et al., Biochem. Biophys. Res. Comm., *17*, 373 (1964)
41. *RIESER*, P., et al., Proc., *116*, 669 (1964)
42. *RONWIN*, E., Canad. J. Biochem. Physiol., *40*, 1725 (1962)
43. *RONWIN*, E., Acta haematol., *26*, 21 (1961)
44. *SCHWICK*, H. J., et al., Z. ges. Inn. Med., *21*, 193 (1966)
45. *SELIGMAN*, A. M., et al., Angiology, *13*, 508 (1962)
46. *SHERRY*, S., et al., J. Biol. Chem., *208*, 85 (1954)
47. *SHERRY*, S., et al., J. Clin. Invest., *33*, 1363 (1954)
48. *SHERRY*, S., et al., Clin. Pharmacol. Therap., *1*, 202 (1960)
49. *SHERRY*, S., et al., J. Lab. Clin. Med., *64*, 145 (1964)
50. *SMYTH*, R. D., et al., Arch. Intern. Pharmacodyn., *136*, 230 (1962)
51. *TROLL*, W., et al., J. Biol. Chem., *208*, 85 (1954)
52. *WOHLMAN*, A., et al., Proc., *119*, 26 (1962)
53. *MEGEL*, H., et al., Arch. Bioch. Biophys., *108*, 193 (1964)
54. *INNERFIELD*, I., et al., Proc., *123*, 871 (1966)

BIOLOGY OF INFLAMMATION

General

Inflammation in its multitude of forms is perhaps the most general and fundamental reaction in all possible pathological processes, for there is hardly any disease which would not include in its course at least some inflammatory secondary phase. Also in the history of evolution, even in life in its early forms, inflammatory signs can be found, e.g. in mollusks or in bone fishes.

The chronological chain of biochemical reactions is, for all inflammations, principally the same, while the symptoms and general picture of events may be quite dissimilar. Deciding factors for these different forms are etiology and localization, also constitution and resistance of the affected organism. Infectious, traumatic, degenerative, allergic, exogenous or endogenous causes set up inflammations in tissues, organs, even involving the whole organism; an acute or chronic course is again an other factor which plays a dominant part in the symptomatology. In spite of the extreme differences between possible locations or causations, e.g. allergy, arthritis, sunburn, hepatitis, neuritis, tendosinovitis, fracture, measles, granulomas, burns, all inflammatory reactions are fundamentally nearly identical in their schematic course. Also chronic inflammations are principally no exception, though the typical symptomatology is often veiled; in some cases it can be recognized only after years by the damage done, like adhesions or tissue degenerations.

Every disturbance of the physiological cell metabolism through chemical (toxic), physical (mechanical) or biological (germs) influences results in inflammation. This is of a complex nature; it involves equally the blood- and lymph-system, connective tissue and more or less the whole organism. The primary seat of inflammation is the mesenchyma, the reticulo-endothelial system. The parenchyma cells and other tissues are only secondarily involved.

At the moment of injury, a series of defense measures starts, apparently with the simple aim of eliminating the harmful damage, or to prevent its further spreading in the organism, by isolating the focus and finally to restore again the original physiology as far as possible. All these defense reactions are grouped under the collective name of inflammatory reactions.

The classical cardinal symptoms of inflammation which date back from the time of humoral pathology but which *Virchow* also accepted, have no importance for the understanding of these reactions; swelling, redness, heat and pain are only noticeable secondary symptoms and give hardly any information about the true nature of the changes taking place.

Only during the recent years research tried to clarify the biochemical processes and their effects. Scientists studied the different phases of inflammation on suitable model reactions and grouped the relatively uniform tissue reactions into a useful scheme. By this means they could not only define and identify the "inflammatory substances" and analyze the inflammatory mechanism, but also determine possible means for our efficient therapy.

The inflammatory processes, as at first useful and important defense reactions, can, however, proceed more vehemently than necessary, so that the organism suffers by its own counter-reactions. There are many examples of such an over-reaction in nature which "shoots beyond its aim". An inflammation should be treated according to severity, extent, cause and state of reactivity of the organism.

Inhibitors of Inflammation

Nowadays a long list of synthetic "inflammation inhibitors" is available which become active at certain points during the course of inflammatory reactions. Their effect is more or less intense and often associated with undesirable side effects. This results in restriction of a longtime therapy which may in some cases entirely prevent their uses. Further restrictions result from the fact that some of these synthetic inhibitors can only be used for very definite specific purposes on account of the mechanism of their action, and others again may have undesirable influence on physiological function, e.g. in form of disturbances of the endocrine equilibrium, of diuresis, of nerve functions etc.

These antiphlogistica are divided mainly into 2 large groups, namely steroids, i.e. cortisone and its derivatives, and non-steroids to which salicylic acid derivatives, butazolidine, indometacine and countless others belong.

The other class of antiflammatory agents consists of enzymatic inflammation inhibitors. These are enzymes which are highly active substances, primarily found in the body itself. They are non-toxic and may be applied in every kind of inflammation. On account of these invaluable qualities, they are not only indicated and desirable for prophylactic application, e.g. sport injuries, but also for long time or continuous therapy, virtually without restrictions.

Their effect and mode of action are for some time firmly established experimentally and clinically. Undesirable side effects are unknown. How logical the use of enzymes as biological inflammation inhibitors is, can be seen from the large amount of physiological and biochemical research devoted to this problem.

Progress and Substances of Inflammation

After an injury starts an inflammation, reactions begin at the point of damage in a succession of a number of physico-chemical events. *Lindner* (8) calls them dissociation of the ground substance.

Very soon after the cell injury disturbances of the permeability of the tissues and in the capillaries are taking place, resulting in the formation of the inflammatory exudate. Experimentally it can be shown that the formation of the exudate is always preceded by the liberation of certain substances from the damaged cells. *Menkin* calls them leucotaxins; he thinks they are protein decomposition products of the cells.

About the chemical nature of the "inflammatory substances" only little is known. While they were regarded in former years as specific inflammatory mediators, to-day they are thought of rather as physiological components of cells which in pathological situations liberate split products with specific action.

The course of an inflammation is morphologically uniform (Uniform Theory), i.e. redness (local hyperemia), swelling and pain are independent of the injurious influences. The uniform course of an inflammatory reaction is supposed to be caused by necrosin, a toxic material, which, according to most recent investigations, is found in damaged cells. Similar albuminous bodies also have been identified in primitive animals like mollusks after cell injuries. Fever as accompanying symptom of inflammation is brought about by a thermostabile polypeptide which is chemically joined to a sugar fraction (Pyrexin).

Closely connected in time with this formation is the process of the body defense in form of phagocytosis. Its fundamental outline was described by *Metschnikoff* 1892 (Phagocytosis Theory of Inflammation). In the area of inflammation leucocytes concentrate, among which cells with polymorphic nuclei, the macrophages, can be distinctly differentiated. Both cell forms are capable of phagocytosis. They take up foreign substances, e.g. germs, fraction of tissues or destroyed cells by engulfing them with their protoplasmic projections. The enclosed particles become covered with a protein-film (opsonin); in the case of bacterial enclosures the bacteriotropin forms this film (*Muld* (12)). A function of this protein film is to reduce the surface tension between leucocyte and the engulfed particle. The extravasation of leucocytes from the intact vessels is brought about by a chemotaxis and is directed by leucotaxin.

About the origin of the macrophages there is no generally accepted theory. Some presume that they also emigrate from the blood vessels, others are of the opinion that they are formed locally by cell division and are derived from vessel wall cells (pericytes).

At the focus of inflammation first polymorphonuclear leucocytes appear, followed later on by the macrophages. This succession is probably caused by the change of pH which can be determined in the inflamed tissues. The hydrogen ion concentration changes first, for a short time, into direction of alkalinity, immediately after to acidity. In this phase the more sensitive polys die, while the more resistant macrophages at first still survive; with increased acidity the latter finally also perish and may form, together with exudate and cell fractions, the pus.

Damaged cells themselves contribute to their destruction. Out of the injured nucleus histones (polypeptides with basic character, rich in lysine and arginine) are liberated, which cause leucocytes to adhere to the vessel endothelium. This again determines the direction for the active emigration of leucocytes.

Besides the local changes, general reactions in the course of an inflammation are also taking place in a number of infectious diseases, like pneumonia, appendicitis etc. The general blood picture shows a more or less marked leucocytosis. Other infections, like typhus, tuberculosis or some virus diseases cause a leucopenia.

Experimentally it can be shown that the inflammatory exudate contains a factor which releases leucocytes (American authors call it the LPF factor). This is a polypeptide joined to an alpha globulin; it causes an increase of leucocytes in the blood stream as well as an increased new formation of leucocytes in the bone marrow. The LPF factor is of great importance in the defense system of the body because the number of leucocytes in the blood determines materially the degree and the duration of an inflammation.

A leucopenia of the blood stream is influenced by two factors known at present as a thermostabile leucopenia substance which eliminates circulating leucocytes and stores them up in different organs, like lungs, liver, spleen. This factor is found in acid exudates while the other, a thermolabile polypeptide with the name of leucopenin is bound to an exudate of basic character.

Fibrin and Inflammation

Fibrin plays a very important part in the process of inflammation. It is formed as soon as fibrogen comes in contact with destroyed, damaged tissue, respectively tissue thrombokinase which is formed from them, or with the peptides mentioned, which at the beginning of the inflammation are formed or liberated. During the fibrin clotting toxic substances are enclosed in the clot. Thus at the beginning of the inflammation a further spreading of toxic products into the organism is prevented. This reaction named "fixation" is already taking place in acute inflammations before the beginning of leucocytosis, and represents an essential biological mechanism, which protects important body organs against a flood of disease-causing agents or toxins, etc. By this means the local reaction becomes an adaptive phenomenon whereby the local negative changes in favor of the protection of vital internal organs are the lesser evil.

The formation of insoluble fibrin leads to a considerable local interference and a stasis of the circulation at the focus of the inflammation. This results in edema and pain. Damage and interference of functions are in the further course eliminated as much as possible by reparative processes. These are primarily helped by proteolytic enzymes of the body, especially plasmin, which liquify the thick tenaceous exudates and depolymerize the fibrin. Even at the beginning of the inflammation those proteolytic enzymes exert this inhibitory function.

During the transformation of fibrinogen to fibrin mentioned before, the tryptic enzymes and plasmin which are present at the same time at the focus of inflammation, exert their inhibitory activity. Biochemically it is an inhibition of polymerization of fibrinogen molecules to micro-fibrin-molecules. Therefore the activity of these proteases consists of liquification by splitting fibrin and other large protein molecules into smaller soluble peptides and amino acids and, besides, in inhibiting the formation of the poorly soluble or insoluble macromolecules present.

In animal experiments it could be shown that the exogenous introduction of proteases before the setting of an inflammation prevents its development altogether or at least reduces it to a mild short irritation. This means that the prophylactic use of tryptic enzymes or papainases stops the development of an inflammation in most cases and practically prevents it. Histochemical tests also prove it. After prophylactic doses of enzymes given 3 to 4 minutes after setting an inflammatory irritation the intercellular and intraarterial fibrin formation is found in markedly reduced amounts in comparison with normal controls.

In screening the literature, it seems strange that little importance has been attributed to the antipolymerization effect of the proteases in inflammatory and degenerative processes. The immediate fibrin deposit is one of the most important defense measures of the organism as it effects an occlusive sealing-up of the damage and thus isolates the focus from the rest of the organism. Besides this safety function, fibrin serves later on as substrate for the regenerative connective tissue cells. Scar tissues or keloid formation or excess deposits of functionless collagen depend to great extent upon formation and local duration of deposited fibrin.

According to *Astrup* (2) fibrin is formed in amounts necessary and sufficient for the healing process. However, problems and at times serious complications happen if excess fibrin quantities are formed and deposited. *Astrup* quotes: "The fibrinolysis is a relatively slow process. Therefore it is to be presumed that the guaranty of resolution of formed fibrin at certain times and under certain circumstances represents a serious problem for the living organism. A delayed dissolution of fibrin may be the cause of a number of pathological processes."

The quantity of the fibrin required for individual purposes depends on the factors of the hemostatic equilibrium like prothrombin, thrombocytes, tissues thrombokinases or fibrinogen. Inhibiting factors of clotting are the proteases, especially plasmin.

A disturbance of the hemostatic equilibrium leading to reduced fibrin formation comprises a series of risks; the inflammation may spread if the isolation is deficient. The wound healing fails or functions unsatisfactorily, respectively with "secondary intention", causing much scar tissue formation. Hemorrhages may appear as deviation of a disturbed blood clotting mechanism. If the hemostatic equilibrium is displaced in the opposite direction, if excess fibrin is formed, which happens more often, it manifests itself in exaggerated inflammatory symptoms: more extensive edema, more pain, complete stoppage of circulation due to compression and microthrombi closing the vessels, a delay of the phagocytosis, delayed healing and increased cell necrosis. If this state continues for some time and the fibrinolysis is delayed or sluggish, large areas necrotize, the healing progresses slowly with excess scar formation. The circulation at the focus deteriorates and the tissue suffers loss of function. Ischemia and increased risk of thrombosis are possible results; fibrin deposits and scars on the arterial endothelium are predilection for plaques and atheroma formation.

Plasmin and Inflammation

The cause of a number of pathological processes seen after acute or chronic inflammation is, according to *Astrup*, an insufficient plasmin activity. While the plasmin niveau of children and young people usually suffices to secure an undisturbed healing process, the plasmin level sinks distinctly in men of about 40 and in preclimacteric women and drops to a low value with old people. This fact explains the increased susceptibility of older people to inflammatory or degenerative diseases, to a more serious course of inflammation and to reduced healing tendency.

Besides the reduction of plasminogen, however, the loss of plasminogen activators in the tissues and of leucocytes is responsible for those handicaps in aging. Also the concentration of the plasmin activator in the individual organs varies a great deal. The highest values were established in the uterus, oviducts, lymph glands, suprarenals, thyroid and prostate, while they are low in liver, spleen, testicles and brain. Correspondingly, organs with high amounts show after injuries only slight scar formation, e.g. uterus, those with low activator contents form heavier scars, e.g. in the liver.

Biogene Amine and Kinines

During the last years a whole new group of substances have been described in connection with inflammation, whose function however, has not been fully clarified, the biogene amines. To this group belong, among others, the tissue hormones, histamine and serotonin which are formed by the decomposition of amino acids and are stored up as granula into the blood and tissue cells. Through cell injury they are liberated from the granula. They lead to an increased permeability of the smallest blood vessels.

The present day opinion is that histamine may possibly have some importance in the early phase of inflammation, but the histaminase present in all cells destroys the liberated histamine promptly. Therefore its activity is limited in time only to minutes or even seconds. The increased permeability by histamine leads to edema on account of the serum extravasation. Preliminary treatment of the organism with antihistamines reduces the histamine level, followed in experimental inflammation by a retarding of edema formation. Antihistamines often have a general body effect of membrane protection.

Serotonin is, like histamine, also liberated from most cells, perhaps also from thrombocytes. It can be found in inflammatory bodies. The name kinine is derived from bradykinine, a member of this group first described and named by *Rocha e Silva*. They are low molecular proteins (polypeptides) which are formed enzymatically from proteins of the globulin fractions of the blood plasma and are called, therefore, plasmakinines. They have a wide distribution in the animal organism. Kinines derived from tissues are tissue kinines (brain, the intestinal lining, also the poison of wasps etc.) During scalding or heating a skin area above 45° C., also kininelike substances, besides histamines, are formed in the serous secretes.

All kinines have a contracting action on smooth muscles. They widen the blood vessels and elevate the permeability of their walls, they also cause pain when they get in contact with the receptor of nerve endings.

The main symptoms of inflammation can be induced by kinines. It is not probable that they have an active function in inflammatory processes. Kinines are not stored up. They are liberated through splitting of a prestage, the kininogens, by proteases. The increased capillary permeability induced by kinines lasts only a short time. Bradykinine activates and accelerates the phagocytic activity of phagocytes.

Phases of Inflammation

Viewed biochemically, the history of every acute or chronic inflammation can be divided into three phases, with very gradual and partly overlapping transitions, which usually prevents any clearcut demarcation. Inflammation is a dynamic occurrence. The findings of modern pathology which attribute great importance to the disturbed homeostasis at a diseased focus, have contributed a great deal to the clarification and understanding of inflammatory reactions. Considering the views of different "schools" on inflammation, the history of an inflammation may be regarded as follows:

In the initial stages, a few minutes after the irritation, a changed permeability of the smallest vessels takes place. A serous or fibrinous exudate is formed which due to pressure alters the state of local circulation. The circulatory slowing up leads to a disturbed homeostasis. The first fibrin nets are formed in the vessels which initiate the screening off of the area. Chemotaxis induced by the products of cell destruction lets leucocytes and macrophages emigrate from the vessels into the focus, where phagocytosis starts immediately. The biogene amines liberated at the same time deteriorate the circulation still further. At the periphery of the inflammatory focus fibrin deposits plug the in- and out-flow of blood by microthrombi within the vessel lumen. This isolating barrier is called "inflammatory membrane". Thus the biological acme of the inflammation has been reached.

Now starts the second phase in which the organism tries to eliminate again the pathological abnormalities and to restore matters ad integrum. The first condition required for this purpose is the restitution of the necessary microcirculation in order to be able to eliminate as rapidly as possible the split-products of decomposition of the high-molecular compounds, which are partly highly toxic.

The improved circulation accomplishes also a better supply of nourishing material and antiinflammatory enzymes like plasmin; also of medicaments to the inflammatory area. In this second phase the leucocytosis is at its height, it also remains for quite some time above normal after the inflammation subsided. This may be regarded as a safety coefficient of the organism for some days. It is a sign that the immunological defense system remains mobilized after cessation of the active signs till it finally returns to normal.

In the third phase, also called regeneration phase, the body attempts to straighten out and repair the tissue damages. Every inflammation leaves a defect, often only microscopically provable, which must be mended by a filling in with connective tissue and new capillaries. The proliferation of new cells is aided by the wound hormones. Another scheme of successive reactions during an inflammation is suggested by G. J. *Martin* (9) which brings the enzymatic activities still more into focus. In his scheme the first very short phase is characterized by the increased permeability of the capillaries. In the subphase, A, lasting only about 2 minutes, histamine is poured out (initial phase of permeability). The following subphase, B, lasts several hours. Inflammation inhibiting amines which tighten the vessel walls now reduce the permeability. The edema causes polymerization of fibrin. In the now following and longer lasting subphase, C, begins an increased histidine formation directed by enzymes, particularly histidine decarboxylases. At the same time bradykinine is liberated from activated kallikrein, and monoamino-oxydase destroys the biogene amines in subphase B.

In the second phase the vessel permeability is reduced still further by the reduction of fibrin, and fluids accumulate in the tissues. Meanwhile histamine is destroyed by diaminooxydases, kinines by kininases.

During the fluid phase fibrin is depolymerized by plasmin, proteins are split into peptones and further down into aminoacids.

The restitution of the circulation, according to *Martin*, is accomplished in the fourth phase, while phase 5 is identical with the regeneration.

Experimental results of the inflammation inhibitors by means of proteolytic enzyme mixtures.

In a comprehensive review of the complex history of an inflammation, the biochemical enzymatic processes are the important aspects in every one of its phases. Therefore it is logical to investigate the possibilities to therapeutically influence them by enzymes, especially proteases. The following describes some experimental research which demonstrates the antiinflammatory action of different proteases, respectively mixture of them, because many experiments have shown that a mixture of different proteolytic enzymes of animal and plant sources have a distinctly superior effect than any of the single enzymes. The clinical applications are discussed in a later chapter.

Innerfield (7) describes in his book "Enzymes in Clinical Medicine" systematic animal experiments demonstrating the antiinflammatory effect of trypsin, chymotrypsin, papain, bromelin, streptokinase and other enzymes. We repeated part of his experiments, using the protease mixture Wobenzyme* and Wobe-Mugos** which we found strongly antiinflammatory.

After subcutaneous injection of egg white into a rat's paw, an

---

* manufacturer: Mucos, 8022 Gruenwald, Germany one dragee Wobenzym contains: 100 mg pancreatin, 40 mg bromelin, 60 mg papain, 10 mg lipase, 10 mg amylase, 24 mg trypsin, 1 mg alpha-chymotrypsin, 50 mg rutin.
** manufacturer: Mucos, 8022 Gruenwald, Germany one dragee Wobe Mugos contains: 25 mg of proteolytic enzymes extracted from pancreas, calf thymus, pisum sativum, lens esculenta and papaya.

inflammatory reaction appears promptly, with marked edema. If Wobenzyme is given orally or rectally or Wobe-Mugos into muscle or into the peritoneum after irritation, the signs of inflammation including edema subside 40 to 70% faster than with the controls. If the same enzymes are given before or at the time of the irritation, the amount of edema is 55 to 75% less. The measurement of the edema was done volumetrically by weight of the amputated paw. Another mode of experimental inflammation is by setting a trauma on the upper part of the leg of a rabbit by a number of standardized hammer blows with resulting edema. The same oral or parenteral treatment before the irritation resulted in a 45 to 70% reduction of edema or hematomas. If the enzymes are given after edema formation, only 2/5 of the time (40%) was sufficient for the disappearance of the inflammatory signs compared to the controls.

Mustard oil applied to a rabbit's eye causes chemosis and conjunctivitis. With pretreated animals the signs are significantly weaker and disappear much faster than with the controls after the same enzyme treatment.

Carbon tetrachloride in rats causes a liver inflammation connected with edema. These are distinctly less with pretreated animals, and if the enzyme is given after injury, the signs disappear in about half the time of that of the controls. Similar were the results with frostbites and with burns, cuts, skin denudation, fractures and other standardized injuries.

As a criterion of antiinflammatory activity the determination of the "spreading factor" can be used. X-ray-irradiated mice receive a subcutaneous injection of Indish Blue. At certain intervals the treated animals are killed and the spread of dye in the skin is checked. With those animals which had the enzyme given before or right after irradiation, the dye spread was between 80 to 240% larger than with the controls. The dye spreads after enzyme therapy easier through inflamed edematous tissue. *Barth and Graebner* (3) describe impressive experiments with mice exposed to X-rays. After total body radiation with 850 r., 98% of the controls died within 30 days, but of those mice which shortly after irradiation had a single dose of Wobe-Mugos into the peritoneum, 50% survived 30 days.

It would lead too far to describe more experiments which all gave similar results to those mentioned. They all prove that proteolytic enzymes are really effective inflammation inhibitors.

Literature:

1. *ABDERHALDEN*, R., Klinische Enzymologie, Georg Thieme Verlag, Stuttgart (1958)
2. *ASTRUP*, T., "Biologie des Plasmins", Schattauer-Verlag, Stuttgart (1967)
3. *BARTH*, G., *GRAEBNER*, H., Dtsch. Med. Forsch. 2 (1963)
4. *EHRICH*, W. W., Mechanisms of Inflammation, Montreal (1953)
5. *EHRICH*, W. W., Die Entzündung in "Handbuch d. allg. Path., Springer-Verlag, Berlin-Gottingen-Heidelberg (1966)
6. *HEISTER*, R., *HOFMAN*, H. F. (editors), "Die Entzündung", Urban & Schwarzenberg, München-Berlin-Wien (1966)
7. *INNERFIELD*, J., Enzymes in Clinical Medicine, McGraw Hill Book Co., New York (1960)
8. *LINDNER*, J., Morphologie, Biochemie und Radiochemie der Entzündung, in "Die Entzündung", Urban & Schwarzenberg, München-Berlin-Wien (1966)
9. *MARTIN*, G. J., Exp. Surg. Res. *202* (1964)
10. *MENKIN*, V., "Dynamics of Inflammation", Macmillan Co., New York (1950)
11. *MENKIN*, V., "Newer Concepts of Inflammation", C. C. Thomas (1950)
12. *MULD*, S., Physiol. Rev., *14*, 210 (1934)
13. *ZINSSER*, H., "Infection and Resistance", Macmillan Co., New York (1923)

GENERAL FIBRINOLYSIS AND THROMBOLYSIS

There is hardly a branch of clinical enzymology which has been, and still is being investigated as intensely and comprehensively as thrombolysis and all questions connected with it. This is equally true regarding fibrinolysis. Between thrombolysis and fibrinolysis there are principally no differences, only due to practical considerations the division was established in the literature. Fibrinolysis takes place under certain conditions in vitro while thrombolysis includes more the complex events of dissolving clotted blood in vivo.

A short historical review shows that the process of blood clotting attracted the attention of scientists for centuries. *Hippocrates* (4th century B.C.) informs in one of his writings that the blood of animals whipped before being sacrificed remains fluid (our modern viewpoint is that great stresses activate the fibrinolytic system to such extent that the blood remained fluid). This phenomenon was rediscovered by *Malpighi* (1648-1694).

Denis (1838) demonstrated that blood clots are unstable. *Dastre* coined in 1893 the word fibrinolysis.

Opie (1911) described proteases in the serum which influence the clotting. From here on there was a long road till the observation of *Tillet* (1933) who demonstrated that in a filtrate of streptococci a fibrinolytic agent exists which is known today under the name streptokinase.

The inactive proenzyme of plasmin (plasminogen) was discovered by *Christensen* (1945).

The induced fibrinolysis for clinical application has been known only in recent years. *Innerfield* (1955) gave trypsin i.v., *Tillet* (1955) also streptokinase i.v., *Cliffton* (1959) plasmin i.v.

Since 1960 a mixture of animal and plant enzymes has been used orally and parenterally by *Wolf* and associates for parenteral and enteric use.

Blood coagulation and fibrinolysis are fundamental conditions for a physiological equilibrium in the human and animal organism and for a smooth course of its biological functions. A normal system of blood coagulation assures the undisturbed circulation of the blood in its vessels and with it the proper nourishment and provision of all body cells and organs as well as the vital removal of the metabolic waste- or end-products. On the other hand, it protects the body against exsanguination after inner or outer injuries.

Disturbances in the course of blood coagulation are the causes of thrombosis and emboli. Thrombo-embolic diseases are familiar to every practicing physician and require immediate therapeutic action to avoid late damages or to save life. A short explanation of the biochemical foundation of blood clotting and fibrinolysis may help the understanding of these processes.

Morawitz, a leading physician in Leipzig, Germany, established in 1905 a scheme of clotting which also is recognized in its outlines today.

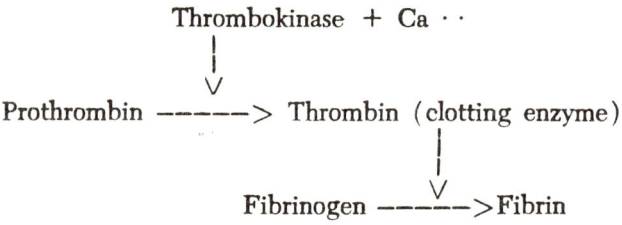

According to *Morawitz* prothrombin is changed to thrombin by thrombokinase. Nowadays we know, however, that this transition of prothrombin is brought about through cooperation of factors present in the blood and others from outside. These factors are called blood thrombokinases and tissue thrombokinases, respectively intrinsic and extrinsic systems.

The newest results of research show that the blood coagulation is a physiological process which is continuously taking place intravasal and which is keeping up the healthy functions of the vessels and their endothelium.

Like numerous other vital functions of the body, also the blood coagulation and its opposite counterpart, the fibrinolysis, underlies a control after the principle of the polar synergism

(*Hoff*). Factors which increase and those which inhibit clotting act singly entirely synergetic, but, when acting side by side, form a synergism of a higher order which is also called dynamic equilibrium between clotting and lysing. Under normal conditions, the coagulation remains in a state of a continual fluctuating balance. The equilibrium is oscillating into the one and the other direction between polymerization and depolymerization. Under pathological conditions in a case of injury, it is one very important local task of the physiologically latent function. According to *Luescher*, the importance of the latent intravasal clotting activity lies in the fact that a fine fibrin film is always deposited on the intima which seals up and protects the endothelium. The coagulation is a complex, very complicated biochemically directed process which functions in single phases and is influenced by about 30 presently known factors which either help or hinder the clotting. The following table gives a summary of the most important blood clotting factors and their usual names:

Factor I Fibrinogen
Factor II Prothrombin (= Thrombogen, Thrombenzym, Prothrombin B, Plasmozym)
Factor III Thromboplastin (= Thrombokinase, Zymoplastin, Cytozyme Thrombokinine)
Factor IV Calcium (= Ca-factor)
Factor V Proaccelerin (= labile factor, plasma Ac-globulin, plasma Ac-factor)
Factor VI Accelerin (= Serum Ac-globulin, prothrombokinase)
Factor VII Proconvertin (= Convertin, stabile factor, serum prothrombin, autothrombin I, Co-factor, Prothrombinogen, Co-Thromboplastin, prothrombin accelerator, prothrombin conversion factor)
Factor VIII Antihemophilic globulin A (= antihemophilic factor AHF, Thromboplastinogen, prothrombokinase, platelet-co-factor, plasmakinin, thrombokatalysin)
Factor IX Antihemophilic globulin B (= Christmas factor, plasma factor X, plasma thromboplastic component (PTC), autoprothrombin II)

Factor X Stuart-Prower factor
Factor XI Plasma-Thromboplastin-antecedent (= PTA, Rosenthal factor)
Factor XII Hagemann factor
Factor XIII Fibrin-stabilizing factor (= FSF, fibrinokinase-factor, Laki-Lóránd factor, LL factor, Hungarian factor).

The names in parenthesis are found frequently, besides others, in the international literature.

Plasma Concentrations:

Factor:	Concentration:
I	200—400 mg/%
II	10—15 mg/%
IIa	
III	
IV	unknown
VII	unknown
VIIa	unknown
VIII	near 10 mg/%
IX	unknown
X	10 mg/%
Xa	unknown
XI	—
XII	1 mg/%
XIII	1—2 mg/%

The fibrinolysis is a later phase of the coagulation in which the fibrin formed during the hemostasis and wound healing is again depolymerized enzymatically (liquified).

It seems especially remarkable that the mechanism of both systems—clotting and lysis—in spite of the opposite functions have a very similar course. The biological activity is bound to proteolytic enzymes, the clotting to thrombin, the fibrinolysis to plasmin.

The physiological connections between both systems are shown in the following simplified scheme (according to the literature):

A simplified scheme of the physiological connections between blood coagulation and fibrinolysis

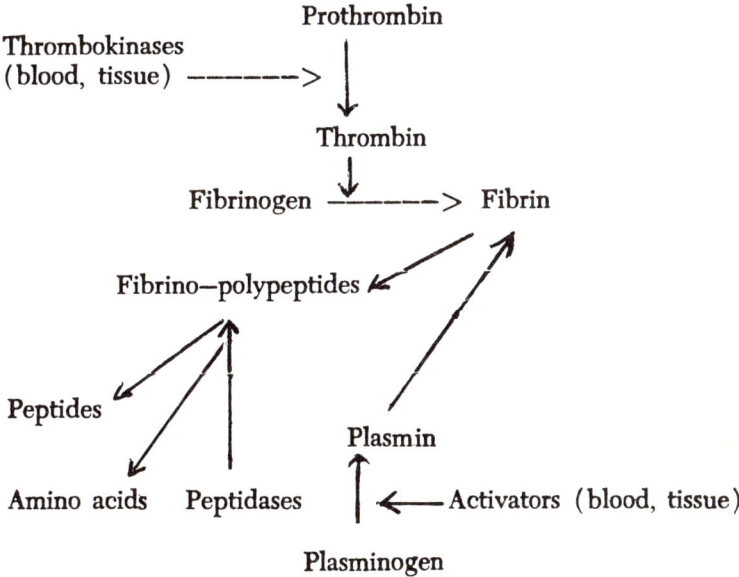

Under the influence of the thrombokinases of which the blood- and the tissue-kinases are better known, an active proteolytic enzyme, thrombin, is formed from the inactive pro-enzyme prothrombin. Thrombin changes fibrinogen by autopolymerization of the fibrinogen monomere into fibrin. This process has been clarified by *Koeppel* in detail by means of electron microscopic investigations.

The transformation of fibrinogen into fibrin can be reduced to four fundamental reactions: adsorption of blood elements to a center of fibrinogen knots, the possibilities of binding fibrinogen threads together, polymerization of the fibrinogen threads which is taking place in three steps, and the influence of the thrombocytes on the course of polymerization and the structure of the formed fibrin.

Conditions for a polymerization are active centers in the compounds which are ready for a combination. Fibrinogen threads have two independent possibilities of combining with each other. One possibility is at the poles of fibrinogen threads and the other at the sides of them (side by side).

At the first polymerization step every time three to eight fibrinogen threads are polar joined to fibrinogen chains which in the second phase are combined by joining laterally to fibrinogen fibers. In a third step of polymerization the fibrinogen fibres combine to fibrin fiber bundles, among these, smaller polymers combine again in larger units.

Thrombin splits enzymatically lower molecular peptides (peptide A and B) from fibrinogen. The fibrin formed is transformed under the influence of the factor XIII in the presence of calcium by the chemical reactions, analyzed only in the last years, into the urea-insoluble fibrin. It appears that a striking parallelism exists in the biochemistry of blood clotting and fibrinolysis. Two of the most important enzymes in the blood are formed in inactive stages for both reactions: prothrombin for the clotting and plasminogen for the lysis.

By a complicated mechanism the transformation of the inactive pre-stage takes place by activators (kinases) into the proteolytic active enzymes thrombin and plasmin.

Enzymes and Coagulation and Fibrinolysis

The investigations of the plasminogen--plasmin system in the serum started in 1898. *Denis* found in serum, treated before with chloroform, a thermolabile proteolytic enzyme. In 1911 *Opie* showed that this enzyme had similar characteristics to trypsin, e.g. the same pH optimum and splitting capacity of casein and gelatin. *Christensen* and *Kaplan* established the following simplified scheme for the plasminogen—plasmin system in the serum:

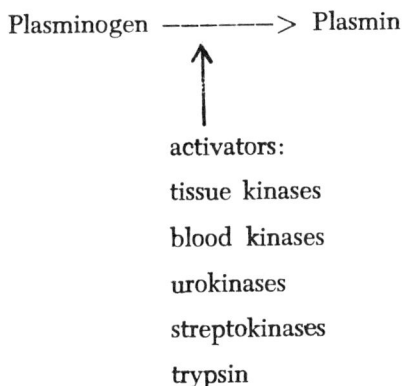

activators:

tissue kinases

blood kinases

urokinases

streptokinases

trypsin

Plasmin and thrombin, the most important enzymes of clotting and lysis also have still other common biochemical characteristics: they are not substrate-specific like other body proteases, they split almost all peptides, and not only at their endgroups.

Still only recently some scientists thought that trypsin and plasmin are identical enzymes. Research with synthetic substrates, however, determined that between those two proteases differences exist in their activator. They proved that trypsin and plasmin are quite different enzymes. The inactive pre-stages trypsinogen and plasminogen have also different activators, and, besides, differ substantially in their property of being blocked by inhibitors like SBI (soybean inhibitor).

The great resemblance of the individual proteases in their physical and biochemical activities led to important clinical conclusions.

Besides plasmin, also other proteolytic enzymes can be used as fibrinolytica, respectively as directly thrombolytic remedies. More details are found in the clinical part of this monograph. To mention only some of the important authors: *Innerfield* and his group describe the positive thrombolytic effects of trypsin and chymotrypsin.

Tillet of streptokinase. *Bein* and other urokinase.

Wolf of animal and plant proteases and their synergistic activity.

Before more detail discussions of plasmin and fibrin, the importance of the latent fibrinolysis should be mentioned.

Importance of Fibrinolysis

Over 100 years ago the peculiar phenomenon of blood was described that formed clots liquefied again. Clotting and fibrinolysis were still recently regarded as processes independent from each other. The main clinical importance of the fibrinolytic process consisted in the dissolving of an already manifest clot. Only since the investigations of *Astrup* (11) it has been known that the fibrinolytic process as physiological reaction in the entire vascular system is necessary for the maintenance of normal biological functions.

As mentioned before, blood clotting and fibrinolysis function are in exactly balanced relation to each other. During the latent physiological clotting of a fibrin film deposited on the arterial intima it is removed again by the also latent fibrinolysis. Therefore in the blood there is always a low fibrinolytic activity present. If this is increased by pathological events, a hemorrhagic diathesis ensues; if reduced, the incident of a sclerotic infarct is favored, according to *Astrup* (11).

In more chronic cases, deposits of cholesterol and lipoids are mostly formed after disturbance of the microcirculation. If this pathological picture is well established, it is called atherosclerosis.

The fibrinolysis is also important for the elimination of secretions and incretions. It assures the patency of the secretory ducts, e.g. of the milk-, tear- or salivary ducts, the respiration and the urinary tracts etc. The fibrin produced during an inflammatory reaction is a preliminary condition for the emigration of histiocytes and with it also for the regeneration process. In such situations the fibrin has accomplished its biological mission after a short time and has to be removed again fibrinolytically. Then the equilibrium moves to the lytic side.

Arteriosclerotic formations of the vessels down to the capillaries can be prevented or at least improved if the intrinsic fibrinolytic potential of the body is raised by exogenous introduction of proteases.

If the fibrinolysis is inhibited or runs below a normal threshold, an increased tendency to scar formation is seen after injuries, like wounds. In case of inflammation of serous membranes, like the pleura or peritoneum, usually adhesions are formed which could be prevented by increased fibrinolysis. Well known also are the complications which are caused by extensive deposits after pleuritis, pericarditis or meningitis.

On the other hand, the formation of abscesses of the lung is caused by the opposite process and can be explained only by a suddenly beginning local fibrinolysis.

Also the physiology of menstruation is directed by a local fibrinolysis, by means of which the clotted matter, which could interfere with elimination, is dissolved in time.

We saw during these considerations that plasmin and fibrin occupy a central position in the fibrinolysis, therefore we may describe them more in detail.

Plasmin

Plasmin is like trypsin, chymotrypsin and pepsin an endopeptidase; it splits off peptide chains from fibrin and fibrin fractions. These peptide chains—polypeptides—are soluble and cannot be coagulated. Fibrin is more sensitive to plasmin than fibrinogen.

The following table shows the most important chemical characteristics of human plasminogen and plasmin (after *Shulman* and others):

	Plasminogen	Plasmin
Sedimentation constant	4.28	3.56
Diffusion constant	2.96	—
Mol-weight	141.000	107.000
Tyrosine	5.1%	6.4%
Tryptophane	2.7%	3.6%
I.P. (iso-electric point)	6.0	7.1

Besides fibrin, plasmin also splits casein, gelatine, beta-lactalbumin, acetylated globulin, serum complement, prothrombin and many others. *Astrup* is of the opinion that all fibrinogen is being polymerized first to intermediate steps of fibrin before it can be split fibrinolytically. For the determination of plasmin the method of *Astrup* has been proved especially suitable: upon standardized fibrin plates the extent of proteolysis is measured.

Antiplasmin

The organism produces a number of antiplasmins which have the function of a regulatory mechanism against the action of proteases. Antiplasmins are able to inactivate plasmin, very rapidly making it inert. Plasmin inhibitors can be found in thrombocytes and serum. Better known and quantitatively determined are the following: Alpha-1-Antitrypsin, Alpha-2-Antiplasmin, Alpha-1-Antichymotrypsin and Inter-Alpha-Trypsin inhibitor.

Inhibitors of human serum

Inhibitor:	concentration Serum, mg/100ml:	mol weight	peptide in %:	carbohydrate in %:
alpha—1—antitrypsin	290.0 + 45.0	54000	86	12.1
alpha—1—antichymotrypsin	48.7 + 6.5	69000	73	24.6
inter—alpha—trypsininhibitor	50.0	160000	90	8.4
antithrombin III	29.0 + 2.9	65000	85	13.4
CI—inactivator	23.5 + 3.0	104000	65	34.7
alpha—2—macroglobulin	260.0 + 70.0	820000	92	7.7

Biochemistry of Fibrinogen

Fibrinogen belongs to the group of globulins. The molecular weight of the beef fibrinogen is 330.000, of the human 340.000. According to older publications, the human plasma contents of fibrinogen are: min. 0,20%, middle value 0,27%, max. 0,36 (*Gram, Lester*).

The amount and constancy of the level of fibrinogen in the blood depend on the proper function of the liver. The destruction of liver parenchyma by sickness, poisons (e.g. carbon tetrachloride) reduces the fibrinogen level which also decreases progressively in hepatectomized animals. *Jonas* found 12-20 hours after hepatectomy a reduction of fibrinogen of 20-50%. With rabbits the drop is still steeper. If only 70% of liver tissue is removed, the fibrinogen level remains constant. The rest of the liver parenchyma is apparently able to form an increased amount of fibrinogen. Furthermore we learn from animal experiments that the organism has no fibrinogen reserves to any extent, and depends therefore on a continuous formation by the liver: food rich in animal albumin raises the fibrinogen level, while hunger, fats and carbohydrates lower it. In inflammations, in some infective diseases, during pregnancy and menstruation the content of fibrinogen in the blood is often considerably abnormal. Some irritations, like changes of the metabolic state or inflammation, stimulate the hepatogenic fibrinogen synthesis.

The rate of metabolism of all plasma proteins is known and can now be determined very exactly by means of isotope technique. Of all proteins in the plasma, fibrinogen has the highest metabolic rate, which also shows the importance of this substance.

The following table shows the half-value times (HVT) for fibrinogen which was marked with different atoms. All measures were figured to the time $t = 12$ hours (after reports in the literature).

Isotope technique:

A tagged with S^{35}

B tagged with J^{131} (Plasma)

C tagged with J^{131} (Fibrinogen)

Authors	Method	HVT	Test animal
Cohen et al.	B	5.6	mouse
Christensen	C	2.7	rabbit
Hammond	C	4.0 – 4.7	man
Lewis	C	4.1 – 6	man
Mc Farlane	C	2.4	dog
Takeda	C	2.5	rabbit
Amris	C	3.2	man

From the half value time it can be figured out that 15-40% of the blood fibrinogen must be metabolized daily. Since the amount of blood fibrinogen under normal conditions is quite constant and no distinct fluctuations appear above or below average, one may conclude that fibrinogen is depolymerized just as fast as it is formed and offered to the organism, and that the fibrinolysis must be in equilibrium with the daily synthesis.

Astrup thinks that in the human body about 2 g of fibrinogen is synthesized; therefore in the same unit 2 g fibrin must be formed and the same amount broken down into split products.

The rapid metabolism of the fibrinogen is surprising; most plasma proteins with smaller molecular weight are metabolized considerably more slowly. The HVT of albumin lies between 15-20 days. An explanation for the preference of fibrinogen in catabolic processes can only be given by considering the total aspect of blood clotting and fibrinolysis. Fibrinogen as a central substance is daily formed at a constant amount and again constantly broken down through a specific enzyme, the plasmin.

With this steady process the connection between clotting and lysis also is clear. An equilibrium exists between both processes: the dynamic hemostatic equilibrium.

The following scheme illustrates this balance and its regulation (*Astrup*).

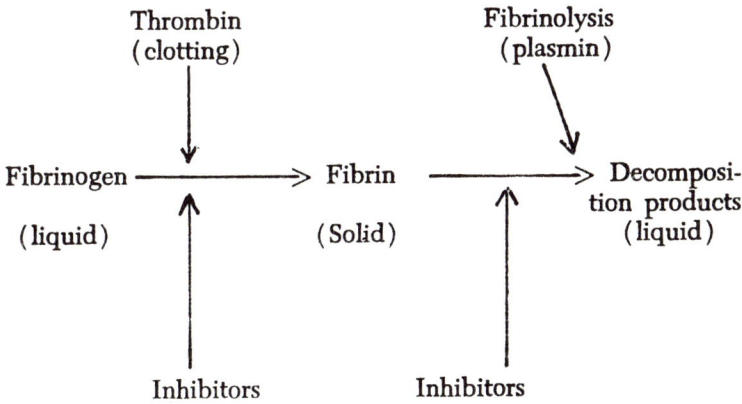

Febrile conditions shorten the HVT 50%, the destruction of fibrinogen is therefore increased. Also during malignant processes, e.g. metastases, the HVT is reduced on account of the higher consumption of fibrinogen.

The metabolic rate of fibrinogen is furthermore dependent on hormonal influences; e.g. it is considerably prolonged in myxedema.

Factor XIII (synonym = Laki-Lorand-factor, LL-factor, fibrinase, fibrin-stabilizing factor).

After being activated by thrombin, it changes the urea-soluble fibrin into urea-insoluble by transglutamination. Therefore the active enzyme is also called transglutaminase.

In conclusion the following scheme shows the proteolytic enzymes which lead to the breakdown of fibrin.

Enzyme in blood

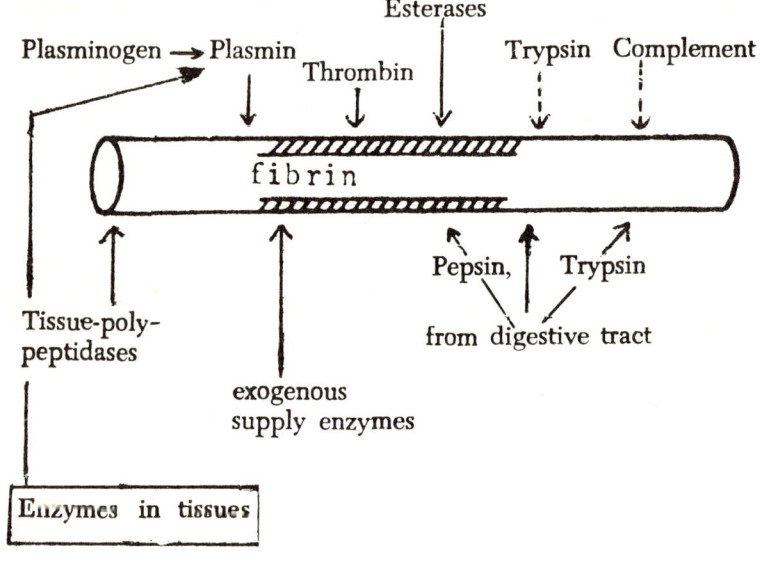

EXPERIMENTAL THROMBOLYSIS

Blood clotting and fibrinolysis, respectively thrombolysis, are found in a normal organism in a balanced antagonism. Reactions governing these processes are regulated enzymatically, as shown before. Fibrinolytic analysis is very difficult on account of the numerous reactions happening simultaneously, its value for treatment of thrombolysis in humans is questionable; clotting, also thrombolysis are influenced by the reaction status of the organism. To this status belong the clotting ability of the blood, the speed of circulation as well as the morphology of the vessel walls and biophysical factors. Pathological processes of inner organs, the presence of infections or inflammatory diseases, damage of the cell oxydation and cell metabolism, and the liberation of toxins after cell decomposition influence the hemostatic equilibrium.

The literature on experimental thrombus formation and thrombolysis is immense—the difference of the fibrinolytic system of the individual kinds of animals compared to man make the judgement of corresponding conditions difficult (*Astrup* (11)). About the end of the last century it was determined that intra-vitam-formed thrombi arise from an aggregate of thrombocytes. *Aschoff* regarded the increase of thrombocytes during a stasis and the increased agglutination of deciding importance for the formation of thrombotic occlusions. *Jürgens* and his group could prove that histamines and toxins which appear in tissues during inflammatory processes increase the tendency to microthrombi formation in the vessel lumen. Also bacterial toxins accelerate the agglutination (*Buchner* (1)).

The wetting time of the vessel endothelium depends on the speed of circulation (*Ritter* (2)). Thrombocytes, being the lightest members of the blood constituents, are normally to be found in larger concentrations close to the periphery of the blood stream. With reduced circulatory speed they are in intimate contact with the endothelium. This results in adherence, clumping and even microthrombi formations. Physiologically caused alterations of the thrombocytes and variations of the electric potential of the vessel wall also favor thrombotic tendency (*Sawyer* (3)). If the vessel wall is damaged by experimental or endogenous inflammatory reactions, it is followed by a secondary agglutination of thrombocytes. Uremic products of metabolism and changes by chronic diseases, including tumors, can also damage the intima in a blood stasis by unknown reasons. *O'Meara* demonstrated in 1967 that all malignant cells secrete a thermolabile prothrombin substance, very similar to Factor III, which very much increases the thrombotic tendency, and is responsible for the so frequent thrombophlebitis and of metastases in malignant disease.

This short summary shows that thrombus formation is a complex and complicated event in which physical, biochemical and pathological factors play an important part. It is to be expected that the process opposite to thrombus formation, the thrombolysis, would have a similarly complicated course. Pathological changes, like trauma, a damaged intima, a slowing up of speed of circulation or chemical changes of the blood constituents move the center of gravity from fibrinolysis towards clotting. By means of a sufficiently activated fibrinolytic potential, however, the normal equilibrium can be established again (*Martin* (4)). The vitally important fibrinolysis shows considerable individual variations. Observation of a day-and-night rhythm of the human fibrinolytic index reveals demonstrable fluctuations, with higher activity during daytime (*Cepelak* (5)). *Biggs* found that physical activity elevates it (39). Plasma lipoids, found in increased amounts in the blood of arterio-sclerotic persons, inhibit fibrinolysis in vivo and in vitro (*Sweet* (6)). During the last months of pregnancy it is much reduced, but immediately after delivery it increases again to the amount before pregnancy started (*Shaper* (7)). We found that stresses have the same effect.

In the first six hours after delivery a distinct increase is demonstrable. Its cause has not yet been explained, but one may presume that the reduced lytic activity during pregnancy is caused by a still unknown inhibitor, which post partum immediately disappears, or there is a connection between the loss of fibrinolytic activity and the increasing level of progesterone and estrogen which stops promptly after delivery of the placenta. In prostate cancer the testosterone raises fibrinolysis which can be inhibited by epsilon-amino-caproic acid (*Rosswick* (8)). Very complicated are the fibrinolytic processes of the newborn which *Edson* (9) thoroughly investigated and which show that in the course of the development the fibrinolytic potential increases. After menopause it decreases steadily till advanced age.

On account of the differences between hemodynamic relation of the extremities, there are also differences in the fibrinolytic activity. *Nilsson* (10) found that arm veins have a higher fibrinolytic potential than leg veins, especially if it is stimulated by an artificial stasis. This difference may explain the generally known fact that thrombi are much rarer in the arm veins with their higher fibrinolytic activity.

Astrup demonstrated that the intima of large veins contains a plasmin activator, but not the arteries. The increased fibrinolytic activity of the blood after venous stasis is possibly caused by liberation of this activator (*Clark* (12)). According to *Ögston* (13) the activity of the plasminogen activators in the venous blood is higher than in the arterial.

In order to be able to imitate pathological processes in the organism, a suitable reaction model must be available. We now know a considerable number of model inflammations. Also regarding thrombolysis there has been much research in producing thrombi experimentally on experimental animals or studying preformed thrombi in vitro. As with many other model reactions, these methods give only an incomplete picture of the events which actually occur in the living organism. They just remain models and no premature conclusions should be drawn regarding the therapeutic value of a remedy from such reactions.

A discussion of the numerous methods of the experimental thrombus formation would be beyond the frame-work of this book. In the appendix are references to published methods of experimental thrombus formations. Depending on this general problem, the investigator will choose the one or the other, being mindful of the fact that these are only model thrombi which are not identical with conditions prevailing in vivo during human thrombosis.

With preformed thrombi which are introduced into a vessel, organization and reactions of it towards the vessel wall can be demonstrated. If for instance, the endothelium is damaged, the thrombus is altered.

In a later chapter it will be shown that foreign cells, e.g. cancer cells, tend to adhere to the endothelial lining, with a loss of its normal structure. These morphological changes are a fundamental condition for the local fixation of cancer cells at the vessel walls and the later formation of metastases. Formation and lysis of a fibrin network about them play an essential role in these processes.

At a focus of hyperplasia of the intima thrombi get quicker formed and organized. The condition of the vessel wall influences this organization and its possible lysis (*Tsapogas* (15)).

Experiments on dogs and cats determined that the organization of a thrombus becomes completed after about 10 days. Therefore results of lysis attempts can only be expected when the therapy begins two or three days after thrombus formation. The findings gathered from animal experiments agree with the experiences made with thrombolytic measures of human thrombosis (*Kamiya* (16)). Since thrombi consist mainly of aggregations of thrombocytes embedded in a network of fibrin, this is readily dissolved by proteases. After several days the fibrin gets replaced by connective tissue (organized) and is then far less accessible to enzymatic lysis. The modern fibrinolytica are enzymes with defined fibrinolytic, proteolytic and esterolytic activity. This suggests an investigation of their activity in vitro on experimentally formed thrombi. The enzymes break down thrombi by proteolysis, they dissolve them. The results of the investigators are very different, depending on the technique and the enzymes used. In the following some research is briefly mentioned: *Kamiya* (16) proved that experimental thrombi are dissolved by 1000 units trypsin or 1100 units urokinase. Also in later states, e.g. after three days, the lysis could be shown histologically. Streptokinase had a stronger activity than trypsin.

Hey (17) describes detailed investigations about the enzymatic solubility of thrombi and of placental infarcts. Frozen sections are covered, by the coverglass method with streptokinase, trypsin, plasmin or chymotrypsin. Tender fibrin nets dissolve in the presence of streptokinase. Older thrombi are attacked only by plasmin, trypsin or chymotrypsin. In order to attack highly organized thrombi, general proteolytic besides fibrinolytic enzymes must be available as the thrombus does not consist only of fibrin.

According to *Hiemeyer* (18) eight hours old thrombi are dissolved by urokinase after two to three hours. The plasminogen concentration drops in the blood, also the fibrinogen level from 250 to 150 mg %. In the experimentally thrombosed ear vein of a rabbit a lysis is brought about after giving trypsin-activated plasmin; simultaneously given heparin increases its effect, (*Andreenko* (19)).

Less favorable are the results which *Sailer* (20) obtained in a similar experimental arrangement with another plasmin preparation. While he was able to produce a microscopically visible lysis, the required doses for complete lysis were not physiological. A total lysis or recanalization was not demonstrable.

Operatively removed thrombi are dissolved in a streptokinase solution (*Ludwig* (21)). The findings are contrary to those of other authors who report that retracted coagula cannot be redissolved by streptokinase (*Schmidt* (22, 23, 24), *Gottlob* (25)). A verification of *Ludwig's* results, whereby the work was done under strictly sterile conditions, revealed that old thrombi are attacked by a solution of streptokinase (*Gottlob* (26, 27)). Further work of this research group demonstrated that fresh, not retracted and plasminogen-rich thrombi, are lysable. Thrombi which have lost their plasminogen supply can only be dissolved when at the same time with the streptokinase plasminogen is supplied. Older thrombi are also lysable without addition of plasminogen by streptokinase, if after two or 3 days a hemolysis is taking place. Possibly in such cases the fibrinolytic compound contained an activator in the erythrocytes. *Künzer* (28) showed the presence of a fibrinolytic enzyme in erythrocytes.

Innerfield (29) was the first to show in animal experiments that thrombi are dissolved by intravenous trypsin. *Sherry* (30, 37) confirms it and concludes from his own experiments: "Undoubtable experimental results prove that in the animal experiment proteolytically active enzymes dissolve arterial and venous occlusions".

Wolf and associates could show that experimental thrombi (on rabbit's ear) are dissolved by giving a mixture of plant and animal proteases; also a recanalization takes place. The passability of the vessel was tested by stab incisions, the flow of blood is proof of thrombolysis and restoration of the circulation. The enzyme mixture is active when given orally, intramuscular and locally.

Thrombolytic activity of enzymes are in interdependence with several locally and generally acting factors. According to *Innerfield*, three possible explanations exist of this process: A complete lysis is possible which however is improbable for individual enzymes; it may take place by a specific mixture of enzymes (broad substrate spectrum); it is also possible that single enzymes bring about a removal of extravascular handicaps, for instance of edema, inflammatory substances, fibrin deposits etc. As a third mechanism a change in the permeability of the thrombi could be considered.

At the bed side, not all, not even most of the occlusions can be reached by therapy. Pure thrombocyte thrombi and so-called mixed thrombi can only exceptionally be attacked therapeutically (*Sandritter* (32)). With increasing age of the thrombus the possibility of lysis is reduced, which can cause considerable difficulties. The usual limit is two to three days during which a thrombolysis is regarded possible, but some American literature extends it to four to five days (*Back* (33)). *Rossolek* (35) could demonstrate still after several more days a lysis of thrombi formed after operations. This agrees with clinical findings of *Gross* and others (34) who saw therapeutically satisfactory lysis after over a week. But in these cases it is hard to decide whether these were genuine recanalizations or were due to the antiinflammatory effect of the enzymes.

According to the research of *Innerfield* and others (36) trypsin inhibits inflammation, *Cohen* (38) proved it for streptokinase. All known fibrinolytic compounds possess a general high proteolytic activity. This makes it probable that their activity on thrombi runs on several lines. It can be readily accepted that besides fibrinolysis, which in its end effect is influenced by antienzymes and physico-chemical conditions of the vessel, also the antiinflammatory and edema-preventing activity contribute to the lysis of the occlusions.

Relatively frequent is the spontaneous lysis of thrombi in vivo. This possibility must always be kept in mind when considering the estimate of thrombolytic activity of medicaments (*Fred* (14)).

The lysis of a clot in vivo has the advantage that the dissolving of the thrombus can be observed macroscopically. It is often difficult to prove lysis in the clinic. A definite certainty can only be established by angiographic tests which however in many cases, as in venous thromboses, will rarely be possible.

For therapeutic thrombolysis the following remedies are available: Streptokinase, Plasmin, Urokinase, Wobe-Mugos (a mixture of plant and animal enzymes).

At present preparations containing plasmin or urokinase cannot be bought in the general trade, also Wobe-Mugos is not yet released by our FDA in the USA.

The following gives a short review about clinical effects.

Some enzyme preparations are sold in the USA as antiinflammatory (Orenzym, Chymoral, Parenzym, Ananase etc.), but their efficiency has not yet been clearly established or objectively investigated.

A thrombolytic therapy can be used in two ways whereby the mechanism of the applied fibrinolytica differs: 1. certain enzymes (lysokinases), to which urokinase and streptokinase belong, activate plasminogen to plasmin; 2. other enzymes like trypsin, chymotrypsin, plasmin, have a directly fibrinolytic and proteolytic activity. The thrombolytic activity of Wobe-Mugos belongs to the second group, whereby, however, the thrombolysis is increased by a wide spectrum of substances as well as by the activation of the body proteases and their precursors.

Literature: Experimental Thrombolysis

1. *BUCHNER*, R., Med. Klin., *61*, 1008 (1966)
2. *RITTER*, A., Thrombose und Embolie, Gruyter, Berlin (1955)
3. *SAWYER*, W. D., et al., Arch. Int. Med., *107*, 247 (1961)
4. *MARTIN*, G. J., Biological Antagonism, Theory of Biological Relativity, Blackston Co., N.Y. (1951)
5. *CEPELAK*, V., Z. Ges. Inn. Med., *21*, 202 (1966)
6. *SWEET*, B., J. Atheroscl. Res., *6*, 359 (1966)
7. *SHAPER*, A. G., et al., Lancet II, 706 (1965)
8. *ROSSWICK*, R. P., Brit. J. Urol., *39*, 143 (1967)
9. *EDSON*, J. R., et al., J. Pediat., *72*, 342 (1968)
10. *NILSSON*, I. M., et al., Lancet II, 127 (1967)

11. *ASTRUP*, T., Fed. Proc., *25*, 42 (1966)
12. *CLARK*, R. L., et al., Angiology, *11*, 367 (1960)
13. *ÖGSTON*, D., et al., Throm. Diath. haemorrh., *16*, 32 (1966)
14. *FRED*, H. L., et al., J. Amer. Med. Ass., *196*, 1137 (1966)
15. *TSAPOGAS*, M. J., et al., Angiology, *17*, 825 (1966)
16. *KAMIYA*, K., et al., Angiology, *12*, 106 (1961)
17. *HEY*, D., et al., Klin. Wschr., *44*, 770 (1966)
18. *HIEMEYER*, V., et al., Thromb. Diath. haemorrh., *17*, 58 (1967)
19. *ANDREENKO*, I., Z. Ges. Inn. Med., *21*, 15 (1966)
20. *SAILER*, S., et al., Klin. Wachr., *41*, 212 (1963)
21. *LUDWIG*, H., Tagung d. dtsch. Ges. f. Angiologie, München (1966)
22. *SCHMIDT*, H. W., Dtsch. Med. Wschr., *88*, 1385 (1963)
23. *SCHMIDT*, H. W., Klin. Wschr., *41*, 1010 (1963)
24. *SCHMIDT*, H. W., Klin Wschr., *42*, 973 (1964)
25. *GOTTLOB*, R., et al., Klin. Med., *19*, 405 (1964)
26. *GOTTLOB*, R., et al., Klin. Med., *20*, 353 (1965)
27. *GOTTLOB*, R., et al., Thromb. Diath. haemorrh., *15*, 570 (1966)
28. *KÜNZER*, W., et al., Klin. Wschr., *41*, 831 (1963)
29. *INNERFIELD*, I., et al., J. Clin. Invest., *31*, 1049 (1952)
30. *SHERRY*, S., et al., J. Lab. a. Clin. Med., *42*, 952 (1953)
31. *JOHNSON*, A. J., et al., J. Exp. Med., *95*, 449 (1952)
32. *SANDRITTER*, W., Behringwerke Mitt., *41*, 37 (1962)
33. *BACK*, N., et al., J. Clin. Invest., *37*, 864 (1958)
34. *GROSS*, R., et al., Dtsch. Med. Wschr., *85*, 2129, 2141 (1960)
35. *ROSSOLEK*, R., Klin. Wschr., *39*, 440 (1961)
36. *INNERFIELD*, I., Ann. N.Y. Acad. Science, *68*, 167 (1958)
37. *SHERRY*, S., et al., Clin. Pharmacol., *1*, 202 (1960)
38. *COHEN*, Sh. G., et al., Circulat. Res., *9*, 851 (1961)
39. *BIGGS*, R., et al., Lancet I., 403 (1947)

THROMBOLYTIC THERAPY WITH STREPTOKINASE

Tillet and associates observed in 1944 that filtrates of certain streptococci cultures are capable of dissolving human clots. The active compound in it is a lysokinase which later on was named streptokinase. It does not directly act on fibrin but transforms the inactive proactivator present in blood into an active form which now activates plasminogen into plasmin.

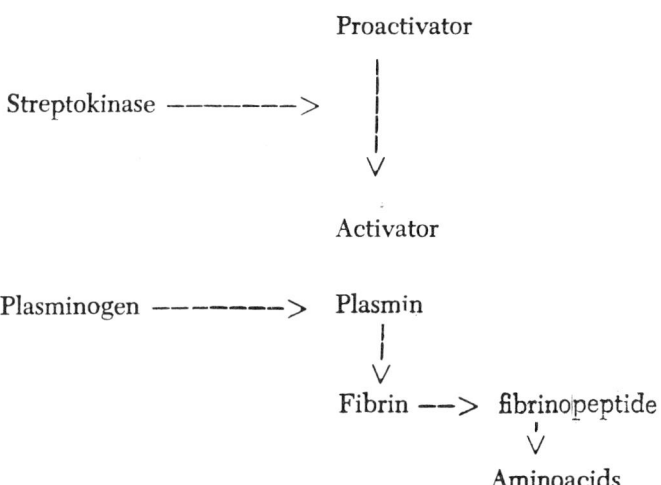

According to Robbins et al. the proactivator-plasminogen molecule consists only of one single polypeptide chain with the amino acids lysine as N- and asparagine as C-terminal substituents. Activation of this molecule with urokinase leads to further two endgroups: N-terminal valine and on the C-end arginine. Therefore on activation the molecule is split into two chains.

Biochemical work pointed out that the activation of plasminogen with streptokinase follows the same way.

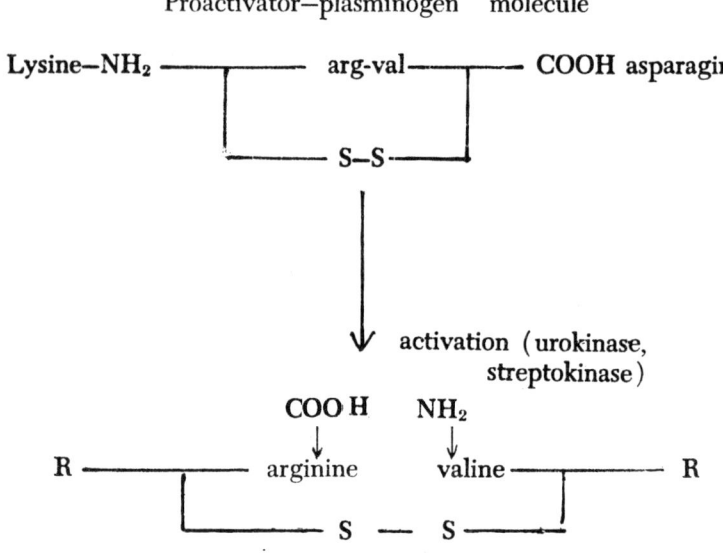

According to K. J. Robbins etc. (J. Biol. Chem. 242, 2333, 1967)

The first preparations commercially available could be given intravenously only by observing several precautionary measures, on account of their poor compatibility. By special purification processes in recent years streptokinase preparations were developed which cause considerably less undesirable side reactions.

Although the human organism has no physiologically present antistreptokinase or inhibitors, yet in every human being there are some immune antibodies against streptokinases. The human blood contains two antistreptokinases. One is an inhibitor which was formed on an immunological basis through past streptococcus infections. Since certainly everybody passed through such infections, this enzyme is almost always demonstrable but its titer underlies very large individual variations. A second non-specific antistreptokinase is described by *Fletcher* (21).

The success of a streptokinase therapy depends therefore on overcoming the antistreptokinase. The maintenance of the therapeutic effect depends on the speed, with which the inhibitors are restored again. Since the streptokinase is rapidly destroyed in the body and plasmin permanently inactivated by relatively large amounts of antiplasmin, a therapeutically sufficient activity in thrombolytic therapy can only be accomplished and sustained by a continuous supply of streptokinase. Many authors ask for a strictly individual dosage whereby the required quantity of streptokinase must be specially determined for each individual patient. Thus, in case of a patient with a high resistance against streptokinase and an insufficient dosage, a hypercoagulability with a progressing thrombosis would result with a "normal" dose; while the same amount in a case of small amounts of inhibitors may constitute an excessive dosage which may lead to increased fibrinolysis and a hemorrhagic diathesis.

The individual doses can be figured out by means of the streptokinase resistance test (*Fletcher*) with the thrombelastograph test of the plasma (*Fischbacher*) or with the analog test of the whole blood (*Fleischhacker* (41)).

In the streptokinase resistance test the smallest amount of streptokinase is determined which lyses the citrated blood of the patient after being clotted by thrombin. The amount thus found figured for the total quantities of blood, gives the required amount of streptokinase which is able to neutralize all antibodies present (TID—titrated initial dosage). This amount is given as a starter dose as a slow intravenous infusion in about 10 minutes and followed up with 2/3 of the amount in hourly intervals as sustenance dose (*Fleischhacker* (41)).

During recent years the determination of TID is mostly given up. At present, after a high initial dose (500,000 to 750,000 U) a dose of 100,000 U streptokinase is given every hour (*Verstraete*) (8)), *Hiemeyer* (17) and others.

With this method a total "plasminogen exhaustion" is established and the danger of hemorrhage is reduced to a minimum by influencing different clotting factors at the same time.

A special problem within the framework of fibrinolysis therapy represents the rethrombosing, for after dissolving of a thrombus, the endothelium, damaged at this spot, remains altered and the danger of renewed thrombus formation at this focus of predilection is especially great. Therefore every fibrinolytic therapy should be followed up with anticoagulantia. The fibrinolytic therapy of a vessel occlusion gives, for the first time, the possibility of dissolving an already present clot; it is a causative therapy.

The decisive criterion for determining a success of the streptokinase therapy is not the recanalization of the occluded vessel but the restitution of its lost functions. But such losses can only be returned to normal if the fibrinolytic therapy leads, before formation of tissue necroses, to a sufficient circulation close to normal. The early beginning of thrombolytic therapy is the essential condition for optimal result.

According to results obtained by clinical application during the last years, the following diseases are suitable for streptokinase therapy.

Fresh (not longer than five days existing) venous or arterial thrombi, lung infarcts and other diseases of the thrombotic complex (*Gross* (2), *Koller* (22), *Fletcher* (21), *Kahn* (24), *Sawyer* (25), *Malmstrom* (23) and others), heart infarctions and certain cerebral infarcts (*Gross* (20), *Kortge* (26), *Poliwoda* (27), *Schmutzler* (28, 29), *van de Loo* (30) etc.)

A success-promising therapy is dependent on certain conditions. The treatment of an acute occlusion must be given within five days (according to others: three days). The collateral circulation must supply a minimum amount of blood, in order that the medicament can reach the thrombus in sufficient amounts; there must not be any special contraindications.

There is a general agreement about the different contraindications among clinicians: General diseases: hemorrhagic diathesis: local tissue defects: disintegrating cancer, fresh operation wound (*Fletcher* (32)), first days after delivery; vessel diseases: arterio sclerosis (*Verstraete* (8)), severe diabetic arterial diseases, endocarditis lenta.

It is not possible to report within the framework of this monograph the extensive literature about streptokinase therapy. The publications mentioned here give some information about the scope of this therapy, its possibilities and limitations. The first favorable experiences with streptokinase therapy of acute heart infarcts were described by *Poliwoda* (27).

Among 12 cases of acute ischemic syndrom of an extremity, the fibrinolytic therapy was seven times successful, the treated patients remained free of recidives after successful anticoagulant therapy (*Hess* (19)).

Winkelmann (12) reports his experiences with thromboses at diverse localizations. Of 23 patients 12 were completely cured and remained free of recurrent clots. Of 36 patients with acute occlusions of arteries of the extremities it was found that the therapy of emboli was much more successful than of fresh thrombosis (*Hiemeyer* (18)). Very good results with emboli were seen in the area of external iliac artery, the popliteal and the arteries of the arm. With occlusions which do not cause severe ischemia or for some reason cannot be operated, the thrombolytic therapy is indicated (*Hiemeyer* (18)). Acute occlusions of the large vessels (int. iliac artery) should be operated.

Patel reported the streptokinase therapy of acute thrombotic or embolic vessel occlusions. He saw good results among 31 patients in 18 cases. Eight were partially successful and five were failures. If an occlusion is older than three days a success is only exceptionally to be expected. There are considerable individual variations in the age of the thrombi and their response to therapy (*Patel* (16), *Hiemeyer* (18)). The following side reactions were seen: fever and leucocytosis the first day, allergic reactions, which could be controlled by cortisone derivatives, nausea and hematomas at the spot of injections.

An important disadvantage of this therapy is the fact that by the antigen effect of the streptokinase a sensibilization (mostly between the twelfth and fourteenth day) ensues. The immune antistreptokinase titer often reaches up to several million units. A second treatment with streptokinase within a year can therefore be performed only in exceptional cases under special precautions (*Johnson* (33), *Witte* (31), *Fletcher* (32)).

Investigations of *Sailer* revealed that the therapy with streptokinase is very promising if it is started early enough. Fever reactions and circulatory disturbances which were often seen before, are of rare occurrence with the now available highly purified preparations, but cannot be entirely avoided (3).

Bross (13) emphasizes that streptokinase for the treatment of massive lung emboli is particularly suitable and may be life saving if the effect is supported at the same time by remedies which facilitate bleeding.

Etiology, location and age of the thrombus influence the results of a thrombolytic therapy. The judgment about success with venous thrombi is difficult, especially if one relies upon measuring of circumferences, skin color or skin temperature. In agreement with the literature (*Schmutzler* (28, 29), *Gross* (20), *Hiemeyer* (17)) found that up to three days old thrombi can be thrombolytically treated with 75% success. Of 76 patients with occlusions of an extremity artery an entire patency of the vessels was accomplished in 36 cases, a partial one in 23. With the other 20 patients the condition remained unchanged in spite of long continued therapy, or they died on the first day of treatment.

With acute cerebral artery occlusions, also with myocardial infarcts, the success of treatment is also depending upon an early lysis of the thrombus, because the tissues are irreversibly damaged through long lasting anoxemia. A sufficient collateral circulation in the area of the cerebral or cardiac occlusion is of course the proper condition for a successful thrombolytic therapy.

Fischer reports his good experiences in the following indications: pelvic venous thromboses, lung emboli, peripheral arterial emboli and thromboses, fresh cerebral and cardiac infarcts. He also found the therapy more successful the earlier it could be started. On account of the antigenic nature of the streptokinase, seven out of 30 patients developed very high fever. The author therefore advised the prophylactic and therapeutic use of corticoids, in order to suppress hyperergic reactions and high temperatures.

Only in the last few years is it possible, by means of the thrombolytic therapy, to treat coronary thromboses within 24 hours after the infarct (*Kahn* (24), *Hess* (19), *Poliwoda* (34) etc.).

Tilsner saw among 56 heart infarct patients treated with streptokinase a distinct reduction of the total lethality including the early death cases. Also the progress of the disease was shortened, in several cases it was possible to restore the working ability earlier than usual. The infarcts were less than six hours old. Only in exceptional cases the treatment was started about twelve hours after the infarct. Since up to now no exact comparative figures between the thrombolytic and anticoagulative therapy have been available the investigations of *Schmutzler* and associates are of special interest.

In a common investigation of six medical clinics of altogether 558 cases of heart infarcts, 297 were treated thrombolytic with streptokinase and 261 exclusively with anticoagulantia. If the treatment started within 12 hours after infarct, the lethality between the second and fortieth day was reduced compared with the usual anticoagulation therapy. The streptokinase therapy does not increase the risk for the patient. The thrombolytic therapy was also not followed by an increased complication rate, compared with the other treatment. The bleeding risk is small. EKG investigations and enzyme determinations in both treatment groups proved that the thrombolytica influenced favorably the EKG. The enzyme activity increases promptly. The reduction of the lethality from 21,7 to 14,1 % is significant.

Hirsh et al. treated 18 patients with acute major pulmonary embolism with streptokinase. Of these, 14 showed clinical improvement. After 24-48 hours of treatment, 12 patients showed definite angiographic improvement. Similar good results have been reported by *Miller* et al., who found satisfactory haemodynamic and arteriographic resolution in 4 patients with acute massive pulmonary embolism.

In femoralis thromboses which rarely (0.5%) happen after diagnostic operations, the streptokinase therapy is especially indicated (*Bostrom* (14)).

Kopp reports about a female patient with massive thromboses of the pelvic veins and sepsis after a varix operation. High doses of streptokinase led to complete cessation of all signs and to improvement of the dangerous disease picture.

Hess (19) warns against the simultaneous use of phenylbutazon preparations and anticoagulantia or fibrinolytically acting remedies, on account of increased danger of hemorrhages.

Several coagulation tests were investigated by *Schneider* during streptokinase therapy, the suitability of some tests for the monitoring of a thrombolytic therapy was determined. The author conducted the following tests before the beginning of therapy, after giving the initial dose, the total dose and 24 hours after beginning of therapy: Thromboplastin time (*Quick*), heat fibrin (*Schulz*), recalcification time (*Howell*) and plasmathrombin time. During treatment he could check up thromboplastin time and heat fibrin. The thromboplastin time increased considerably, the recalcification time reduced slightly. The authors conclude from their findings that a single test for the monitoring of streptokinase therapy is insufficient and recommend the methods of *Schulz* and *Quick*.

About the specificity of the antistreptokinase titer report *Baumer* et al. and describe a special precipitation technique by means of which the specificity can be increased.

Literature: Streptokinase

1. *SCHMUTZLER*, R., Dtsch. Med. Wschr., *91*, 581 (1966)
2. *OHLER*, W. G. A., Med. Klin., *61*, 140 (1966)
3. *SAILER*, S., Fortschr. Med., *80*, 775 (1962)
4. *FLEISCHHACKER*, H., et al., Med. Klin., *57*, 703 (1962)
5. *OHLER*, W. G. A., Med. Klin., *59*, 61 (1964)
6. *POLIWODA*, H., Angiology, *17*, 528 (1966)
7. *KOPP*, P. H., Münchn. Med. Wschr., *109*, 193 (1967)
8. *VERSTRAETE*, M., et al., Brit. Med. J., *1*, 675 (1964)
9. *SAILER*, S., et al., Wien Med. Wschr., *118*, 73 (1968)
10. *SCHNEIDER*, K. W., et al., Blut., *12*, 275 (1966)
11. *BÄUMER*, A., et al., Z. Rheumaforsch., *26*, 78 (1967)
12. *WINCKELMANN*, G., et al., Scand. J. Lab. Invest. Suppl., *16*, 7 (1964) and *16*, 15 (1964)
13. *BROSS*, W., et al., Med. Klin., *62*, 755 (1967)
14. *BOSTRÖM*, H., et al., Läkartidn., *63*, 420 (1966)
15. *OHLER*, W. G. A., Med. Klin., *61*, 140 (1966)
16. *PATEL*, J., et al., Presse Med., *75*, 589 (1967)
17. *HIEMEYER*, V., Dtsch. Med. Wschr., 92, 955 (1967)
18. *HIEMEYER*, V., et al., Med. Klin., *60*, 583 (1965)
19. *HESS*, H., et alm., Med. Klin., *60*, 1208 (1965)
20. *GROSS*, R., et al., Dtsch. Med. Wschr., *85*, 2129 (1960)
21. *FLETCHER*, A. P., et al., J. Clin. Invest., *33*, III (1959)
22. *KOLLER*, F., Dtsch. Med. Wschr., *90*, 1233 (1960)
23. *MALMSTRÖM*, G., et al., Proc. 8th. Congress Europ. Soc. Hematol., II, 433 (1961)
24. *KAHN*, P., et al., Wien. Klin. Wschr., *73*, 677 (1961)
25. *SAWYER*, W. D., et al., Arch. Int. Med., *107*, 274 (1961)
26. *KORTGE*, P., 60 Tagung der Nordrhein-Westfälischen Ges. Inn. Med., (1963)
27. *POLIWODA*, H., Vorh. Dtsch. Ges. Inn. Med., Wiesbaden (1963) and Angiology *17*, 528 (1966)
28. *SCHMUTZLER*, R., V., Hamburger Symp. über Blutgerinnung (1962)

29. *SCHMUTZLER*, R., Experimentelle and therapeutische Fibrinolyse, Schattauer, Stuttgart, p. 143 (1963)
30. *van de LOO*, J., et al., Verh. Dtsch. Ges. Inn. Med., Wiesbaden (1963)
31. *WITTE*, S., Behringwerke Mitteilungen, *44*, 33 (1964)
32. *FLETCHER*, A. P., et al., J. Clin. Invest., *38*, 1096 (1959)
33. *JOHNSON*, A., et al., Throm. Diath. haemorrh., *5*, 391 (1964)
34. *POLIWODA*, H., et al., Dtsch. Med. Wschr., *91*, 973 (1966)
35. *KÖRTGE*, P., et al., Dtsch. Med. Wschr., *92*, 1546 (1967)

THROMBOLYTIC THERAPY WITH PLASMIN

The biochemical and pharmacological characteristics of plasmin have been discussed earlier. Plasmin is successfully given in diseases of the blood vessels (thrombo-embolic occlusions). From the multitude of publications only a few may be mentioned.

Popkin (1) gives a clinical summary of the methods for treating peripheral occlusive diseases so far used. Besides cortisone derivates and vasodilators, he also points to plasmin with which he had encouraging results; however, the small number of cases does not allow a final judgement about this therapy.

Moser (2) reports about the application of plasmin in acute deep thrombophlebitis. The local injection into or next to the thrombus is probably superior to its systemic use. The author could not give a definite statement about the value of this therapy (1961).

Sheffer (3) brings a survey on his experiences of three years with the use of plasmin in thrombo-embolism. 67 patients with thrombophlebitis, lung emboli and retinal thrombi received plasmin in high doses. With all groups the results were satisfactory. Deciding is the clinical impression, the in vitro results obtained are better than the clinical. The author emphasizes that further experiences with plasmin therapy must be collected.

Five female patients with severest edemas and venous occlusions after breast operations were treated with plasmin. Although the author cannot give information about the mechanism of the effect, he regards a recanalization as certain (*Evans* (4)).

One of the first communications about the plasmin therapy originated from *Clifton* (5, 7) who treated 40 patients with severe primary diseases (cancer) besides their thrombophleboses. In the majority of cases the fibrinolytic effect in the blood was elevated, tested on the euglobulin lysis time. During therapy the swellings, edemas and tension partly subsided, as well as pain. In a few cases the thrombus could not be found by palpation.

A lysis of the thrombus could be determined with certainty only on two patients with pulmonary emboli. The time of lysis may vary between 15 minutes and 48 hours. The often considerable side effects of the therapy are certainly caused by the streptokinase content in the preparations (added as activators).

Ambrus (6) confirms on his own patients' material the findings on *Clifton's* and stresses the point that plasmin, of all proteolytic enzymes so far investigated, had accomplished the best therapeutic results. Desirable are still investigations which may lead to further purification of the preparations. Central venous occlusions of the retina respond very well to plasmin therapy if given very early and in high doses. Also encouraging are the successes in circulatory disturbances which are connected with glaucoma (*Drance* (8)).

Dechtjar (9) thinks he noticed an analgesic effect with repeated infusions of 22 thrombotic patients who were simultaneously treated with plasmin and heparin. 13 were substantially improved, 8 only moderately; without success was the therapy of a patient suffering with ischemic foot gangrene.

Literature:

1. *POPKIN*, R. J., Angiology, *12*, 427 (1961)
2. *MOSER*, K. M., et al., Angiology, *12*, 1951 (1961)
3. *SHEFFER*, A. L., et al., Angiology, *12*, 165 (1961)
4. *EVANS*, J. A., Angiology, *12*, 155 (1961)
5. *CLIFFTON*, E. E., Ann. N.Y. Acad. of Science, *68*, 209 (1957)
6. *AMBRUS*, J. L., et al., Ann. N. Y. Acad. of Science, *68*, 97 (1957)
7. *CLIFFTON*, E. E., et al., Circulation, *14*, 919 (1956)
8. *DRANCE*, S. M., Angiology, *12*, 149 (1961)
9. *DECHTJAR*, A. L., et al., Klin. Med. Moskau, *45*, 88 (1967)

THROMBOLYTIC THERAPY WITH UROKINASE

Source and importance of urokinase isolated from urine were still a matter of disagreement a few years ago. Only newer investigations showed that urokinase is a strong plasminogen activator and dissolves thrombi in vivo and vitro. (*Johnson* (1), *Williams* (2)). Human urine contains besides urokinase (*Astrup* (7), *Celander* (8)) also substances which promote clotting (*Frey* (4), *v. Kaulla* (5, 6)).

The place of the primary formation of urokinase is not fully ascertained. It is under discussion whether the urokinase is derived from the epithelium of the eliminating ducts of the urine or found in the blood and then excreted in the urine (*Ladehoff* (9), *Bjerrehuus* (10)). In the latter case the urokinase would be an excretion form of the plasma profibrinolysin activator.

Different authors showed that the urokinase is formed in the kidney, mainly in the juxta-glomerular cells and is secreted from there into blood and urine (*Buluk* (11), *Januscko* (12), *Prokipowicz* (13), *Woronski* (14)). The urokinase is active in a pH range of 4 to 10 (*Celander* (8)).

Kropp et al. investigated to what extent changes in the excretion of urokinase occur with certain kidney diseases, or whether secretion is depending on interferences of renal functions. The test was done by means of the method described by *Hartert* (15), with TEG, which is specially suitable for substances with high activity (*Astrup* (16)). In TEG the course of fibrinolysis caused by urokinase has a typical curve; a nearly plateau form and fairly fast narrowing of the curve (*Astrup* (16)). The study of many TEG curves after giving fibrinolytically active compounds showed that the fibrinolysis acts not only after the end of the fibrin coagulation, but also during the clotting process (*Astrup* (16)). In kidney diseases the excretion of urokinase does not depend on the type of the disease and is therefore not suitable for differential diagnosis (*Smyrniotis* (17)).

There exist, however, connections between the urokinase activity and the kidney functions. In general, the excretion of urokinase is lowered by reduced kidney function, also in uremia the urine contains only very small amounts of it (*Smyrniotis* (17)).

The urokinase elimination is independent of the concentration of urine. It shows no diurnal variations, no differences between male and female or between different age groups (*Smyrinotis* (17)).

The level of urokinase excretion is therefore quite a good measure for the kidney function. Several authors presume that urokinase plays an important part in the prevention of clotting processes in the urine, principally of the formations of cylinders (*Frey* (4), *Matsumara* (18)).

An equilibrium exists between coagulation promoting and inhibiting influences of which the urokinase represents only one factor in the urine.

A trial to apply urokinase therapeutically suggested itself because it was to be expected that the compound, a normal body substance, does not have antigenic characteristics (*Sherry* (25), *Hashimoto* (22), *Hansen* (23)).

The production of highly purified preparations is, however, difficult and the separation of the fibrinolytic activity from thromboplastic acting combinations (urothromboplasmin in urine) is not completely successful (*Caldwell* (24)).

Newer investigations of *Raab* (26) proved that urokinase has some slight antigenic characteristics. Anaphylactic reactions could be increased in experimental animals by giving them urokinase.

According to *Fletcher* (20) the toxicity of intravenously given urokinase is slight and the blood clotting system is only little influenced. 32 patients with thromboembolic vessel disease received infusions with high doses of urokinase (250,000 U) up to 31 times per patient. The thrombolytic activity was several 100% higher with the treated patients than with the controls. The author does not give any success figures but stresses the necessity of a broad clinical test of urokinase in blood vessel diseases.

Since urokinase preparations for the clinical use are available only in very few countries, the literature about its application is scarce. More detailed data were given in 1967 by *Sautter* (21) in his work about the use of urokinase in acute lung emboli. In such cases a rapid lysis of the thrombus is especially important, because otherwise a prompt embolectomy must be considered. The medicament was applied either intravenously or through a catheter introduced into the pulmonary artery. The clinical results were excellent with eight patients and good with two. If fully restored to normal, with three the findings were good, the findings were based on arteriography, three patients were while among the remaining four patients no worthwhile changes could be observed. With three patients who were already prepared for embolectomy, after use of the medicament a dramatic improvement took place.

In agreement with the findings of the streptokinase therapy the authors showed that fresh arterial thrombi are lysed faster than lung thromboses. Side effects were never seen.

Literature: Urokinase

1. *JOHNSON*, A. J., J. Clin. Invest., *42*, 145 (1963)
2. *WILLIAMS*, J. R. B., Brit. J. Exptl. Pathol., *32*, 530 (1951)
3. *KROPP*, R., et al., Med. Klin., *62*, 1941 (1967)
4. *FREY*, J., et al., Med. Klin., *59*, 560 (1964)
5. *KAULLA*, K. N. v., J. Lab. Clin. Med., *44*, 944 (1954)
6. *KAULLA*, K. N. v., Acta Haematol., *16*, 315 (1956)
7. *ASTRUP*, T., et al., Proc., *81*, 675 (1952)
8. *CELANDER*, D. R., et al., Amer. J. Cardiol., *6*, 409 (1960)
9. *LADEHOFF*, A. A., Scand. J. Clin. Lab. Invest., *12*, 136 (1960)
10. *BJERREHUUS*, I., Scand. J. Clin. Lab. Invest., *4*, 179 (1952)
11. *BULUK*, K., et al., Experientia, *18*, 146 (1962)
12. *JANUSUKO*, T., et al., Thromb. Diath. haemorrh, *45*, 554 (1966)
13. *PROKIPOWICZ*, J., et al., Thromb. Diath. haemorrh., *12*, 396 (1964)
14. *WORONSKI*, K., et al., Thromb. Diath. haemorrh., *12*, 87 (1964)
15. *HARTERT*, H., Z. exp. Med., *117*, 189 (1951)
16. *ASTRUP*, T., et al., Amer. J. Physiol., *209*, 84 (1965)
17. *SMYRNIOTIS*, F. E., et al., Thromb. Diath. haemorrh., *3*, 257 (1959)
18. *MATSUMURA*, T., et al., Experientia, *22*, 318 (1966)

19. *HANSEN*, P. F., et al., Angiology, *12*, 367 (1961)
20. *FLETCHER*, A. P., et al., J. Lab. Clin. Med., *65*, 713 (1965)
21. *SAUTTER*, R. D., et al., J. Amer. Med. Ass., *202*, 215 (1967)
22. *HASHIMOTO*, I., et al., The Green Cross Corp., Osaka, 19 (1967)
23. *HANSEN*, P. F., et al., Angiology, *12*, 367 (1961)
24. *CALDWELL*, M. J., et al., Thromb. Diath. haemorrh., *9*, 53 (1963)
25. *SHERRY*, S., M. *FISCHER*, Int. J. Clin. Pharmacol., *2*, 165 (1967)
26. *RAAB*, W., et al., Z. Biol., *115*, 310 (1966)

CLINICAL RESULTS OF ENZYME THERAPY IN INFLAMMATIONS AND THROMBOSES

In conclusion, if the clinical work with streptokinase, urokinase and plasmin is critically evaluated, we may state that the therapy with streptokinase for certain indications may promise success, but is bound to steady controls and monitoring investigations by a clinical laboratory. About dosage for different indications there is no final scheme at present.

Therefore it was to be expected that the interest of the clinician has been directed to the proteolytic and fibrinolytic-acting enzymes of the own body which were on the market.

In numerous publications American authors reported about the use of highly purified enzymes of the pancreas, e.g. trypsin for the treatment of peripheral and thrombo-embolic blood vessel disease (*Innerfield* (6, 7, 8, 9), *Laufman* (14), *Moser* (16), *Jenkins* (11), *Tillet* (19), *Hardy* (5), *Eddy* (1), *Kryles* (13), *Tagnon* (18)). *Innerfield* was the first to call attention to the successful use of trypsin on 18 patients suffering from thrombophlebitis.

Laufman gave 30 patients with acute and 11 patients with chronic thrombophlebitis trypsin in form of continuous drip infusion, without any noticeable side effects. Within the first three days after beginning treatment a definite improvement started in all acute cases, with reduction of accessory inflammations. Among the chronic cases the swelling and tension reduced and the pain disappeared.

Fisher et al. gave trypsin intramuscular or intravenous to 135 thrombophlebitis cases. The acute cases showed a favorable reaction after 12 hours: pain and swelling subsided, also local redness. The blood sedimentation rate became normal, also the temperature. In spite of the relatively small enzyme dose (10-50 mg i.v. or 2,5 to 5 mg i.m.) the effects were prompt. No undesirable side effects appeared, only slight pains at site of injections could sometimes not be avoided. The authors added to their publications the information that they were able to clinically cure by above treatment 75 to 80% patients with acute thrombophlebitis. Of the chronic progressing cases the results of the treatment were not quite so favorable.

In conclusion, the authors point to the fact that intramuscularly applied trypsin is no substitute for a therapy with anticoagulantia, but represents an independent form of therapy with which the course of acute thrombophlebitis can certainly be shortened (3, 4).

By their epicutan-test *Fisher* et al. investigated the sensitiveness of the patients to trypsin and came to the conclusion that doses above 50 mg trypsin should be avoided. Since all patients received an anti-histamine therapy before the trypsin injections, this method of therapy is restricted.

Probably the largest experience in the therapy with proteolytic enzymes of thrombo-embolic diseases the working group about *Innerfield* (9, 10) collected. He surveys the therapeutic results of far more than 1000 patients with blood vessel diseases and asserts that the intramuscular application of trypsin in acute superficial and deep thrombophlebitis shortens the duration of the disease considerably and makes an ambulant therapy possible, which often results, in a short time, in a full restitution of working ability of the patient.

In the last years much research was done to improve the enzyme preparations or to use a mixture or combination of different enzymes with different substrate specificity. Publications of several authors appeared in the literature about the use and effects of a mixture of animal and plant proteases (Wobe) on inflammatory diseases or vascular occlusions. *Valls-Serra* (20) describes in detail the anti-inflammatory and anti-edematous action of the balanced mixture of animal and plant enzymes, contained in the preparations Wobe-Mugos and Wobenzym and discusses thoroughly the clinical results which he gathered on patient material of 245 cases. The investigations were restricted to different groups of patients with whom, on account of preliminary trials, a decided and quick effect was to be expected. The following table gives the types of indications:

Endarteriitis obliterans	15
Thrombangitis	6
Raynaud disease	7
Scleroderma	4
Acrocyanosis	5
Vasculitis	10
Arterial Thromboses (peripheral and cerebral)	25
Arterial Embolism	6
Deep Thrombophlebitis (acute)	26
(subacute)	43
(chronic)	63
Superficial Thrombophlebitis	4
Lymph Edema	8
	222

In acute cases 2 to 3 injections were given daily, all others received only one daily. In acute arterial thrombi, the author combines the enzyme therapy with anticoagulantia and continues after a few days with the enzyme mixture alone. With this routine therapy it was remarkable how much the reconvalescence was shortened and the number of recurrences reduced.

The preparation was also given prophylactically in arterial surgery. Also here the results were very good. In contrast to the therapy with anticoagulantia, there was no hemorrhagic risk. The following table gives the results of the enzyme therapy in treating superficial and deep thrombophlebitis:

Number and types of cases		Results			
		excellent	satis-factory	%	slight
acute	26	18	6	92	2
subacute	43	22	15	86	6
chronic	63	33	21	85	9
edema with and without ulcerations					
superficial	23	16	5	91	2

In 85 to 92% the therapeutic results were excellent. This can probably be explained by the synergic activities of the fibrinolytic, antiinflammatory and edema reducing effect of the mixture.

The following medical histories taken from *Valls-Serra's* article give a good example of the therapeutic possibilities connected with enzyme therapy: 37 year old patient with hypotonia and hypercoagulability of the blood. When he was 20 years old he recovered from a pleurisy complicated with phlebitis of the left leg. Since then this leg remained edematous. Now after 16 years an indolent ulcus formed on the lower third of the leg. At his first visit a very extensive edema of the left leg was found with insufficient connections between the vena saphena and the femoralis. The ulcus was infected and very painful. After eliminating the infection a phlebo-extraction of the saphenous vein was performed. The edema subsided very rapidly and the ulcus healed. Some time after discharge the patient developed a thrombophlebitis of the other leg. The therapy consisted only in giving heparin (the enzyme mixture was not available at that time). There was only a partial improvement, the bilateral edema which led to a recurrence of the old ulcus of the left leg remained unchanged. Only now he had a chance to use the proteolytic enzyme mixture. As a result the edema subsided rapidly without returning any more. Independently of it and in spite of the subsidence of the edema a new, more distally located ulcus near the ankle of the same leg was caused by a trauma as often happens in a postphlebitic scleroderma. Again the proteolytic enzyme mixture was applied, though it was dubious whether after such a long period since the beginning of the original phlebitis an effect could be expected. However, the ulcus healed promptly and the edema disappeared.

The very good results obtained by *Valls-Serra* are gaining weight by the fact that he used control tests straight along during his clinical work (also lympho- and phlebography).

Wolf reports on more than 400 carefully observed patients treated by the protease mixture for venous diseases. The individual cases were very carefully observed, followed up and compared with controls. He used the proteases preferably in form of suppositories. The list of patients and results of therapy are summarized in the following table.

It is to be added that the majority of chronic cases had been treated before, some for months, locally also with antibiotica or anticoagulantia. The desired success only took place by the enzyme therapy. Also, practically all patients mentioned subsidence of pain 24 hours after the beginning of therapy. The edemas lessened or disappeared and temperatures returned to normal.

Anticoagulantia and other treatment methods used before may be continued, if desired, during the enzyme therapy, but as a rule they are superfluous. The enzymatic thrombolysis does not lead to any complications, like embolus formation. The thrombus fixed to the vessel wall is being dissolved from the centrum out to the periphery (wall) of the vessel and does not form any embolus.

Wolf found also that chronic thrombophlebitis cases responded promptly to enzyme therapy, but had the tendency to recur after interruption which, however, could be controlled again by resumption of enzyme doses.

	simple phlebitis	superficial phlebitis	deep phlebitis	with complications, ulcus etc.	with marked edema
Number of cases	96	123	15	64	49
Age (average)	19–81 (59)	24–79 (58)	31–64 (53)	47–72 (62)	37–67 (51)
suppositories given	15	20	30	35	36
Results: very good	24	71	10	30	18
good	60	36	4	27	21
slight or poor	12	16	1	7	10

Abstract of Fantoni's table of cases under Wobenzym therapy.

Case	Name	Age	Diagnosis	Operated	Enzyme-therapy	Cured	Days	Side reactions Incompatible
34	O.E.	62	Abscess	Operation	1 Amp./die	−	−	−
35	S.L.	61	Canal fistula	Operation	2 Amp./die	+	−	−
36	B.A.	66	Gluteal Abscess	Operation	1 Amp./die	+	−	−
37	T.E.	32	Perianal Abscess	Operation	1 Amp./die	+	9	−
38	S.E.	40	Rectal Abscess	Operation	1 Amp./die	+	−	−
39	C.A.	58	Thrombophl.	Phlebotomy	1 Amp./die	+	5	−
40	B.A.	58	Thrombophl.	Phlebotomy	1 Amp./die	+	−	−
41	B.A.	59	Thrombophl.	Phlebotomy	1 Amp./die	+	−	−
42	C.A.	58	Varicoplastic	Phlebotomy	1 Amp./die	+	11	−
43	I.L.	51	Varicophlebitis	Phlebotomy	1 Amp./die	−	−	−
44	C.R.	75	Thrombophlebitis	Phlebotomy	1 Amp./die	+	−	−
45	B.E.	44	Thrombophlebitis	Phlebotomy	1 Amp./die	+	15	−
46	C.C.	69	Varicophlebitis	Phlebotomy	1 Amp./die	+	15	−
47	V.D.	62	Orchiepididymitis	Phlebotomy	1 Amp./die	+	15	−
48	B.R.	42	Orchiepididymitis	Phlebotomy	1 Amp./die	+	10	−
49	B.A.	53	Orchiepididymitis	Phlebotomy	1 Amp./die	+	10	−

In another research study *Fantoni* (2) reports about the clinical test of the same multiple enzyme preparation as biological inflammation inhibitor. In his elaborate tabulary summary of cases in general surgery, also of several patients with inflammatory, pus-filled infiltrations of varices and thrombophlebitis, the influence upon the edema and the course of wound healing supplied the criteria determining the effect of the therapy. Among all patients with blood vessel diseases the complaints subsided surprisingly fast and the duration of the sickness was shortened.

At a colloquium about the possibilities of enzyme therapy many scientists from Czechoslovakia and Germany communicated their experiences with the enzyme mixture therapy in August 1968. *Lisicky* (15) used the preparation on patients with thrombophlebitis or phlebothromboses. They were either postoperative complications or independently starting diseases. Two or three days after beginning therapy all complaints eased off and edemas subsided. There were hardly any failures. *Stasek* (17) of the Central Oncological Institute of the Charles University in Prague reports about the influencing of phlebothromboses by intramuscular and orally applied Wobe-Mugos. His report shows an unequivocal general success, though the relatively small number of cases (21) does not permit a final judgement about the therapeutic merits. The results were similarly prompt.

Kaderabek (12) used the preparation on patients with post-thrombotic syndromes of the lower extremities. The previous therapy (rest, elastic bandages, vasodilator remedies etc.) brought no relief and the patients activities aggravated their condition, as they were farmers and consequently could not adhere to the instructions. Only after beginning of the enzyme therapy the symptoms subsided promptly and most patients soon recovered.

Wolf and others reported on some cases of apoplexy, infarcts and lung emboli which are success-promising indications for enzyme therapy.

The advantages connected with the therapeutic use of enzymes must not be overlooked: a reliable inhibition of inflammation, with fibrinolytic effect, leads to prompt and lasting elimination of pain and subsidence of edema, without danger of hemorrhage.

There are no important undesirable side reactions, and tedious laboratory controls of the clotting status are not required. However, one rare side reaction should be mentioned. In agreement with most authors only in extremely rare cases allergic symptoms appear after parenteral therapy, which can be controlled with certainty by corticoids or antihistamine medication. In very rare cases the treatment had to be discontinued.

To avoid a tedious enumeration of all the many authors of whom we have received reports, we just mention as representative *Varo* (Düsseldorf) who had more than 350 cases. He uses the enzyme mixture as routinely successful therapy in thrombophlebitis or phlebothrombosis of different localizations. He states that fresh cases of thrombophlebitis react within 3 days concerning dolor, rubor, color and functio laesa.

Complete remission is usually achieved in 2-3 weeks of treatment. Some severe cases required a higher dose and prolonged treatment.

THERAPEUTIC USES OF PROTEOLYTIC ENZYME MIXTURE IN DIFFERENT INFLAMMATIONS

From the above discussion one might get the impression that the use of the proteases may be restricted to inflammations of blood vessels. However, the field of usefulness is considerably wider and the statement is justified that every inflammation may be an indication for enzyme therapy. It would fill a whole volume for a comprehensive discussion of diseases for which enzyme therapy may be useful. All enzyme therapy is antiinflammatory, and inflammation happens to be the central process of most diseases.

Innerfield's monograph (26) gave some insight into these therapeutic possibilities. In the meantime our knowledge of enzymes has been considerably increased. We know that the antagonism of enzymes against inflammation is a logical therapy; we can foretell in which diseases enzyme application would be probably successful. In the following no dosages or systematic plans of therapy will be reported; rather some individual indications, chosen at random, which helped to collect experiences. Large doses are mostly desirable to "overplay" the inhibitors formed; there is no upper limit to the enzyme dosage.

One particularly successful sphere of application of enzyme therapy is in sport lesions, respectively accidental traumas. The consequences of luxations, contusions, compressions or lacerations can be cut short or prevented, hematomas and pain removed.

High doses of Wobe-Mugos or Wobenzym are given after such injuries. In one case, immediately after a severe fall, 17,5 g enzyme (70 tablets) were given upon the wish of that patient. The patient had landed full force upon the left side of face which injury was bound to lead to marked edema and hemorrhage. However, except for a small cut at the eyebrow, the next day no visible sign was left, only the conjunctival vessels were slightly injected. Another patient who fell down a whole flight of stairs received four hours after the fall 30 tablets Wobenzyme and Wobe salve locally to the ankle joint, which meanwhile was very much swollen, discolored and painful. The next day all signs had disappeared, she could go to work without discomfort. Similar rapid enzyme effects are reported by *Lichtmann* (28) with prompt edema reduction. *Fulgrave* (23) communicates his experiences with sport injuries. In many cases the usual forced inactivity for a two month's rest period after major injury could be reduced to two weeks. Also sprains and strains to knees or muscles can be impressively controlled by proteases. In some states of the Union prize-fighters are directed to take enzymes prophylactically before the fight in order to prevent before-hand the results of severe injuries.

In general protease doses taken prophylactically offer several advantages to sportsmen exposed to injuries (boxing, wrestling, football, baseball etc.), without suspicion of so-called "doping". Also bursitis, tendosinovitis, "tennis elbow" etc. are prevented or helped by it, according to reports of sport physicians, such injuries prove often resistant to other therapies for long periods.

Several physicians noticed excellent, at least temporary, results in ischialgia. A patient with ischias recurring regularly twice a month has been without attack now over a year while taking daily four tablets prophylactically. After interrupting the therapy for three weeks the symptoms returned, but disappeared immediately after resumption of the enzymes. Other diseases of sensory nerves, bones or cartilage respond similarly to the therapy. Fractures heal much faster and with less complications. Hematomas and bruises disappear in a short time. Even an extensive decubitus of a patient who had to stay in bed for long periods, healed within 10 days. In general dirty or abscess wounds very soon show a clean floor without pus when enzyme salve with or without antibiotics is applied.

In some gangrenous processes the progress can be stopped by local and systemic enzyme therapy. *Innerfield* was able to eliminate diabetic gangrene with it.

Similarly the healing process of furuncles, carbuncles, deep abscesses or fistulas is considerably hastened by proteases. The same experiences are seen in cases of even extensive burns of 2nd and 3rd degree. Skin transplantations show, after a preliminary enzyme therapy, a distinctly increased healing rate; the same is true with acid or alkaline burns. The necrosed part is cast off soon and with less scar formation.

Female diseases: mastopathies are favorably influenced by enzyme salve application, combined with tablets (*Lautz*). Impressive examples are also adnexitis, oophoritis, salpingitis, metritis, parametritis etc. Patients regularly report a prompt cessation of complaints. Acute conditions show mostly fast improvements while chronic diseases need continued therapy. For epistotomies the protease therapy has been used post-operatively for a long time; always a prompt healing tendency, edema reduction and absence of stitch abscesses are seen, facts which *Fulgrave* also observed. In cases of portio erosions, bougies or suppositories of enzymes are introduced into the vagina. Whether the elimination of vaginal fluor can be accomplished exclusively by local enzyme treatment is not certain. It stopped in a number of cases.

Urogenital Tract

Several specialists used proteases in diseases of the kidney and urinary ducts and reported very good results in the majority of cases. There are three reports on patients with urolithiasis, who remained free of complaints for many months. The urine frequently shows the elimination of "sand" during this treatment.

In kidney diseases the fibrinolytic activity of the enzyme mixture and disappearance of casts seems to be the main effect since it keeps open the kidney cevelicula. There are many reports about benefits of enzyme therapy with inflammation of testicles or epididymitis (*Fantoni, Miller,* et al. etc.), even those after parotitis.

Also the enzyme therapy of prostatitis is very helpful, the flow of urine becomes free.

Diseases of Air Passages and Lungs

Besides the antimicrobial therapy, proteases prove most successful in diseases of the respiratory tract. The fibrinolytic and mucolytic effect leads to a rapid liquefaction of tenaceous secretions and exudates, which can be easily coughed up or otherwise eliminated. The continuous irritation by inflammatory products adhering to the respiratory lining delays the recovery of the epithelial cells. Also the effect of any medicines used is weakened by being diluted with respiratory secretions. Laryngitis, bronchitis or pneumonias are helped, the inflamed lining returns to normal sooner.

In cases of lung abscesses or empyema, an injection of the enzyme mixture directly into the abscess cavity is indicated if possible. One to four ampoules dissolved in physiological salt solution are injected. There are also favorable reports on protease therapy of emphysema.

It appears that also the formation of some silicoses could be reduced with sufficient amounts of proteases, according to preliminary reports on investigations started on a broad scale a year ago.

In cases of pleuritis (*Franz*) and pleura effusions the use of enzymes is indicated. Adhesions and heavy fibrous plaques are largely prevented. As prophylactic measure the enzyme mixture proved its value with older people bedridden on account of other diseases, who were in danger of congestive bronchitis or pneumonia due to forced immobility. The experiences with aerosol therapy with a few patients were excellent.

A strange case of aspiration pneumonia may be mentioned. The patient developed a therapy-resistant bronchitis with pus excretion for over ½ year. After about one week of protease therapy he coughed up a small partly macerated chicken bone apparently swallowed or inhaled before the cough started. The symptoms promptly stopped.

Grimminger reports on his experiences with the enzyme therapy in diseases of lungs, bronchi and pleura, resp. postoperative complications after thorax operations. He emphasizes that only a well tolerated enteral and parenteral available combination of animal and plant enzyme (like Wobe-Mugos) allows a broader use in different affections of the thorax organs. The preparation accomplishes the following in thorax diseases:

1) Solution of microthrombi in the area of the diseased parenchyma, combined with a better blood circulation and increase of the level in the tissues of antibiotica and chemotherapeutica given at the same time.

2) Activation of lymph absorption.
3) Enzymatic depolymerization of fresh fibrin coagula in the alveoli.
4) Rapid resorption of inflammatory edemas, detumescence of the bronchial mucous membranes, improvement of the blood and lymph circulation.
5) Gentle liquefaction of the bronchial secretion and of the empyema pus by intrapleural application.
6) Rapid improvement of the general well-being.

In the treatment of unspecific parenchyma processes, e.g. lobar-, segment- and broncho-pneumonias and of virus pneumonia one starts with a combined broadband-antibioticum plus enzyme therapy, whereby the choice of the antibioticum depends on the antibiogram. Specific processes and Candida infections have to be treated specifically. In case of early application of enzyme a rapid reduction of the disease symptoms and a complete restitution can be expected. A delayed application cannot prevent the beginning cornification and fibrosis, but can reduce them to a minimum.

In cases of lung abscesses and abscess-forming pneumonias the enzyme mixture is given orally. In case of larger abscess cavities, twice a week instillation therapy with the antibioticum combined with 1 to 4 ampouls of Wobe-Mugos is performed, thus a rapid cleaning up of the abscess wall is accomplished.

In case of early start of the enzyme therapy, an infarct or an infarct pneumonia already in the stage of organization can be influenced to a great extent up to a complete remission. In bronchitis of different etiologies the oral enzyme therapy leads to fast detumescence of the bronchial mucous membrane, to a liquefaction of the secret and with it to an easing of expectoration. The frequency of cough is reduced. In sero-fibrinous or suppurative pleuritis the instillation of the enzyme mixture with an anti-bioticum has been very successful. In cases of early application the formation of fibrin plaques can be mostly prevented. In the thorax surgery the enzyme mixture has been of great value for post-operative use in cases of chronic empyema with pleurectomy and decortication as well as with lung resections. By the accelerated dissolution of fibrin and hematomas and the increased resorption of exudates and wound edemas the cure is enhanced, the X-ray picture and the cosmetic result of the operation scar is improved. Also the tendency to post-operative thromboses is greatly reduced.

Adelberger and *Wörn* investigated the effectiveness of enzyme administration per rectum in different kinds of pulmonary complications, such as: specific empyema after pneumolysis, extrapleural empyema, former cauterization and other states. They report valuable improvements. (See also *Adelberger* etc. in the chapter: Vitamin A.)

Upper Respiratory Tract

Inflammatory condition of throat, nose, mouth and ear are particularly unpleasant to the patient on account of pain and discomfort. The usual "painkillers" frequently do not bring desired relief. In such cases the proteases in candy form are useful. Indications are sinusitis with its postnasal discharges, rhinitis, otitis, laryngitis, pharyngitis etc., also postoperative edemas after septum-tonsil-eardrum operations etc. One candy is usually taken every hour to reduce swelling and pains. Similar treatment after eye operations is helpful. Inhalation therapy with enzyme has just been started.

Dentistry

The therapy is indicated in all inflammations about the teeth and the mouth cavity; also after extractions and root operations; the therapy with protease candies is preferred.

Plastic surgeons who have been using the protease mixture for years prophylactically and therapeutically are praising the great help of this medication in their work. The healing proceeds faster, discomfort is eliminated and the resulting scar lines are almost invisible.

Above examples show that the desirable use of proteases is nearly unlimited. The inhibition of inflammation by the harmless protease enhance the biological efforts of nature. The expression "inhibitors" would be more correctly replaced by "accelerators" of inflammatory processes, for protease do not "stagnate" inflammation, they hasten the successive stages and thus shorten the duration of the disease. Proteases help the elements of the body defenses (homeostasis) to more actively concentrate on the elimination of the inflammatory products and pathological substances of the body.

Literature

1. *EDDY*, V. M., J. Kansas M. Soc., *18*, 464 (1955)
2. *FANTONI*, P., Clinical testing of Wobe-Mugos as biological inflammatory inhibitor in general surgery. Lecture before the Med. Enzymforsch. Ges., Grunwald, July 29, *68*.
3. *FISHER*, M. M., et al., Angiology, *8*, 60 (1957)
4. *FISHER*, M. M., et al., N.Y. J. Med., *54*, 659 (1954)
5. *HARDY*, E. G., et al., Surg.-Gynec. & Obs., *100*, 91 (1955)
6. *INNERFIELD*, I., J.A.M.A., *156*, 1056 (1954)
7. *INNERFIELD*, I., Surgery, *36*, 1090 (1954)
8. *INNERFIELD*, I., Clin. Res. Proc., *2*, 35 (1954)
9. *INNERFIELD*, I., et al., J.A.M.A., *152*, 597 (1953)
10. *INNERFIELD*, I., et al., J. Clin. Invest., *31*, 1049 (1955)
11. *JENKINS*, B., J.M.A. Georgia, *45*, 431 (1956)
12. *KADERABEK*, F., Pribram (CSR), Personal Information (1968)
13. *KRYLES*, L., et al., Ann. New York Acad. Sc., *68*, 178 (1957)
14. *LAUFMANN*, H., et al., A.M.A. Arch. Surg., *66*, 552 (1953)
15. *LISICKY*, J., personal information (1967 and 1968)
16. *MOSER*, K. M., New England J. Med., *256*, 303 (1957)
17. *STASEK*, V., personal information (1967 and 1968)
18. *TAGNON*, H. J., Proc. Soc. Exper. Biol. Med. *57*, 45 (1944)
19. *TILLET*, W. S., Harvey Lect., ser., *45*, 149 (1952)
20. *VALLS-SERRA*, J., Medizina Clinica, Barcelona, number 4 (1967)
21. *WOLF*, M., information from the Biological Research Institute (New York)
22. *WOLF*, M., Arch. Enfermed. Coraz. Vas., *69*, 114 (1966)
23. *FULGRAVE*, E. A., Ann. New York Acad. Sc., *68*, 192 (1957)
24. *FRANZ*, S., Esslingen, personal information
25. *HOFFMAN*, G., Rastatt (1967)

26. *INNERFIELD*, I., Enzymes in Clinical Medicine. McGraw-Hill Book Company, New York (1960)
27. *LAUTZ*, H. J., New-Ulm (1967), personal information
28. *LICHTMANN*, A. L., Ann. New York Acad. Sc., *68*, 196 (1957)
29. *MILLER*, J. M., et al., Mil. Med., *118*, 31 (1956)
30. *GRIMMINGER*, A., Erfahrungsheilkunde number 1, 18, 1971

SUBSTITUTION THERAPY BY DIGESTIVE ENZYMES

The substitution therapy for correction of a congenital or acquired enzyme deficiency of certain organs, like the pancreas, has already been in use for many years, whereby the required raise of the enzyme level is brought about by exogenous supply of the corresponding enzyme.

A complete absence of pancreatic enzymes is extremely rare, but the physician frequently meets cases connected with deficient enzyme production. Certain disease symptoms point to the fact that the enzymes splitting or digesting proteins, fats or carbohydrates are produced in insufficient amounts or prematurely become inert. Both result in insufficient digestion of the meals or of individual food products.

Deficiency states resulting from diminished enzyme production are seen in a great number of general, but also of specific diseases affecting pancreas, liver and digestive tract.

Digestants containing pepsin and hydrochloric acid have been in use for many years for the treatment of stomach disorders and belong to the earliest therapeutically useful enzyme preparations. It has a favorable influence on hypo-acidity of the stomach, the acidity supplying the optimum pH for pepsin activity.

For the substitution therapy of pancreatic insufficiency, the 3 groups of enzymes mentioned are of special importance. A great number of such preparations containing these enzymes are on the market. The principal constituent of such mixtures is mostly pancreas powder which a priori contains protein-, fat- and carbohydrate-splitting enzymes in physiological combination. Some also contain bile salts and acids or their derivatives, also pepsin combinations.

Such very specific-acting enzymes enable the physician to relieve symptoms in the one or other desirable directions, e.g. accelerate sluggish digestion of certain food substances; prevent bloat and other symptoms of indigestion etc. Sometimes orally given polyenzyme preparations must be protected against partial or complete destruction in the stomach by its pepsin and hydrochloric acid by a special coating. The manufacturer is even able to determine the exact area in the digestive tract where the enzyme is released. Also the time in relation to mealtime is often figured out so that the active digestive enzyme meets the food at the proper time and place. The digestive effect of pancreatic enzymes is often enhanced by the addition of corresponding plant enzymes (*Klücken* (3)). Reliability regarding dosage and effect is only possible by standardization of the individual enzymes (*Bamann* (1), *Dirr* (2), *Merten* (4)).

For the substitution therapy with enzymes there is a long list of indications. We do not wish to go, within this frame of discussions, into the broad topic of digestive and metabolic disease or the different indications and preparations available. Just a few random examples may give an idea of these therapeutic possibilities:

Insulin therapy of pancreatic diabetes is the classical example of substitution therapy and is supposed to compensate for an enzyme deficiency or to establish again an essential equilibrium; therefore its application is especially desirable when a deficient secretion is found.

"Indigestion" comprises diseases of liver, gall bladder, pancreas, the digestive tract proper, and many other disorders with digestive symptoms. Some of them are closely connected with enzyme deficiencies, certain dyspepsies, meteorism, *Roemheld* complex, fermentation and putrefaction dyspepsias etc. Whether connected with enzyme deficiencies or not, most of them are helped by enzyme medication.

It is natural that enzyme medication is helpful after operations on the pancreas or during chronic pancreatitis, for weak digestion in old age or for digestive disturbances due to certain modern medicaments or types of foods common nowadays.

Prophylactically, enzyme preparations are helpful with large rich meals or hard-digestible foodstuffs.

A prolonged regular intake of pancreas enzymes (pancreatin) preparations with the main meals may have a beneficial influence on the function of the pancreas.

Indigestion due to greasy foods is common and usually caused by dysfunction of the gallbladder or liver, but frequently it is due to a deficient pancreatic function. The different pancreatic preparations available, some fortified by fungus lipases, prevent post-prandial discomfort or gallbladder attacks.

Bloat, feeling of fullness or cardiac discomfort after meals, if not caused by nervous habit of aerophagy, is mostly brought on by fermentation of carbohydrates and can usually be prevented or relieved by preparations containing amylases which are mostly manufactured from fungal extracts. Pepsin and pancreatin (uncoated) are not given together as digestants, because their pH optimum is so far apart that they inhibit each other.

Literature

1. *BAMANN*, E. etc., Arzneim.-Forsch. (Drug. Res.), **4**, 35 (1954)
2. *DIRR*, W., Munchn. Med. Wschr., 1936, p. 1750
3. *KLÜCKEN*, M., etc., Z. klin. Med., *153*, 527 (1956)
4. *MERTEN*, H., Dtsch. Z. Verdauungskr., *10*, 159 (1950)

AGING AND ENZYMES

In recent years numerous connections between natural and pathological aging could be clarified. Thereby it has been found that enzymes are of very considerable importance in gerontology and geriatrics. In advancing aging enzymes are formed in reduced amounts and partly reduced activity. As the following discussion will show, these facts are partly responsible for the development of the symptoms of aging and also for premature aging.

In old age and the aging processes men showed an active interest already since early antiquity. The principal interest was mainly concerned with possibilities and means of preventing by any means the aging process and its consequences. Authors of antiquity, but also of modern times up to our days have expressed in big volumes their theories and hypothesis about the aging complex. There has been also no lack of phantastic, mystic and hardly realizable advices how to make a younger man out of an old one.

But only in our most recent time the problems referring to old age and the aging person were put on a scientific basis. While also still today numerous opinions and theories exist of the aging process of the organism, the "Fountain of Youth" tendency, to make an older man young again, has nowadays only historical interest. Gerontology is the science of changes during aging. It is a special branch of medicine with a steadily increasing importance. Gerontology and Geriatry (diseases of old age) are dealing not so much with problems of prolonging life and to develop prophylactic measures against them. What importance this branch of science really has, the following discussions will indicate.

During the past 40 years, thanks to advances in medicine, the average span of life has been considerably lengthened. But this does not apply to the maximal age limits which apparently have hardly changed since thousands of years; nowadays just many more people than before approach this maximal span of life.

For instance, the life expectation in USA was about the turn of the century 47.3 years, in 1950 it was already 68.2 years, therefore 21 years longer. If this tendency would progress in a straight line, the theoretical average age limit about the year 2000 would be about 90 years. The actual prolongation of the average life expectancy since 1900 has apparently self limiting causes, like better hygiene, healthier mode of nutrition, better possibilities of differential diagnosis, fight against and control of epidemics, of infectious diseases in general and better control of infant mortality. A conspicuous reduction of mortality figures took place about 1940 with the introduction of the antibiotica. Since then the number of old people, the "senior citizens", form a steadily increasing part of the population. The number of people above 65 years of age in the USA multiplied and at present consists of over 15 millions, a full twelfth of the total population. It has been shown by different authors that the mortality rate of man forms an exact logarithmic curve with age.

Primitively organized beings underly special regulations regarding aging. Microorganisms, also viruses, apparently multiply continuously without showing any visible or otherwise determinable sign of aging. Although during their growth in artificial culture media degenerative aging processes are taking place after many generations which resemble aging processes of higher organized beings, these are not identical with them.

Leucocytes show aging by morphological changes, by segmentation of their nucleus. They cannot divide anymore and perish after a certain time. The gain of function acquired by higher differentiation of the individual cell is paid for by the loss of ability to divide, and the death of the cell. An epithelial skin cell is still able to divide at the basal layer and progresses gradually to the upper layers. This is connected with a higher differentiation. The epithelial cell forms enzymes or mucous substances; after some time it is cast off or dissolved by enzymes. This constitutes death of the individual cell but for the organ whose function has to be kept up, it is the basis of a continual regeneration. While the degenerated or mutated cells are dissolved by autolysis, i.e. by the enzymes contained in the lysosomes, the cellular organism, the organ, remains fully functioning. If however such useless cells remain within the cellular organism, on account of lack of a catabolic enzyme or a weakness of the small sensitized lymphocytes containing cellular antibodies, the particular organ will show symptoms of aging. This will be more pronounced, the more such cells without function are remaining in the tissues. The activity of their enzymes during aging is more and more reduced which would explain the progressive loss of function caused by aging. Therefore *Astrup* (1) sees in the diminution of the enzymes, especially of the plasmin, a parameter for the biological aging.

Cells having an especially vitally important function cannot divide anymore, on account of their high degree of differentiation. Their function has to last without weakening during the entire life, as for instance heart muscle fibres or ganglia cells. If a high percentage of them becomes disfunctioning through mutation, it is bound to lead to irreparable signs of aging or senility; a regeneration e.g. by stimulation of the rate of mitoses is not possible anymore.

Besides these aging processes of the individual cells, a loss of function due to age develops in tissues or organs by deterioration of their supplies of blood and lymph system. The vessels and the intercellular substances are especially important, because upon their exact supply depends the vitality and function of all parenchymal cells. Also their age-dependent structural changes are decisive for the signs and symptoms of aging. The higher differentiation of such tissues and their dependence upon the source of nutrition by intact vessels and mesenchyma as well as their control by the nervous system, however, is responsible also for their increased susceptibility to aging processes. Therefore not only the cells are primarily aging, but also the mechanism of their nutrition and the intermediate substance, for instance the connective tissue and, depending on it, the whole tissue complex.

The numerous theories trying to explain the aging processes are beyond the framework of enzymes, here we may only discuss a few briefly.

The somatic mutation theory as an important cause of aging could be supported very recently by experiments with ionizing radiation by *Curtis* and *Szillard* (6). Spontaneous mutations with and without mitosis are taking place continuously in all tissues available to investigation, and they accumulate with age. Mutated cells are found in large numbers in almost all organs of old men as well as old animals. These are abnormal cells and they reduce the functions of the corresponding organ. By means of mutagenic ionizing rays one can produce identical results like they are seen during natural aging; that means one can induce an accelerated aging process. A quantitative determination of mutations is very difficult because they can be identified only during mitosis of normal or also abnormal division of chromosomes. Certain cells (e.g. bone marrow, intestinal lining) divide so rapidly that the derivatives of cells with abnormal chromosomes (chromosomal breaks etc.) perish, after several divisions. They are weaker, biologically speaking, than normal cells. Young animals treated with ionizing rays remain seemingly healthy over long periods. But aging signs start prematurely with them and they die earlier, due to the ray-induced chromosomal damage. Investigations showed that short living animals have normally already a higher mutation rate than long living ones.

Cells with frequent mitosis, like bone marrow or skin, are very ray-sensitive and have a certain tendency to malignant degeneration. Brain cells cannot divide, are little ray-sensitive and degenerate only exceptionally. Changes due to aging are hardly observed on nerve cells. Apparently a great number of neurons is present in inactive state and has no physiological function. In brain surgery it is known that large areas of such "silent" brain tissue can be destroyed or removed without deficiency symptoms. It has been figured out that in every human being from early life on about 10,000 brain cells perish daily without being replaced by mitoses. Reserve cells take the place of their function; also all functions remain fully intact as long as such reserve cells are still available. If they are used up, irreversible deficiency signs appear.

Interesting are the consequences of mutations of the sex cells. In the ovaries at the beginning of sexual maturity all ova, in immature form, are present which can be fertilized during the following 30 to 40 years. Due to mutagenic influences the number of altered ova increases steadily. This explains the fact that older mothers give birth more often to offsprings with retardation caused by mutations, e.g. mongolism. With mothers 45 years old the probability of birth of a mongoloid child is more than 20 times higher than with those of 25 years. With fathers this is quite different. A man, 60 or 80 years old, hardly produces more retarded children than one of 20 years. Spermatozoa are steadily formed, different from ova, in extremely higher amounts, and continuously. Defective or mutated sperm cells have less vitality for swimming and therefore cannot reach the mature ovum in the oviduct in competition with healthy vigorous germ cells. This natural qualitative selection among the spermatozoa is an example for the mechanism of evolution.

The proteolytic enzymes are regulating the control of the aging process due to mutations. The cell membrane and cytoplasma of healthy cells contain sufficient inhibitors against the endogenous proteases, which could many times be shown in cell cultures. Mutated cells, however, show a reduced amount of inhibitors, therefore a selective susceptibility of these mutated cells toward endogenous proteases exists. Thus the organism is protected against the formation and accumulation of defective tissues; at the same time mutated cells are also attacked by cellular antibodies.

In the blood and lymph of young people a large amount of proteases is present which suffices to destroy to a great extent mutated cells present. In advanced age, however, the proteolytic potential of the serum gets more and more reduced; at the same time the amount of inhibitors increases, thus the mutated cells survive to a great extent and are able to multiply. For this reason multiple pathological conditions develop, characterized as diseases of old age. Also the much increased frequency of cancer in advancing age can be explained by these facts.

If the organism is supplied for a long time, best during the entire life, beginning at a certain age, e.g. 40 years, with proteolytic enzymes, it stands to reason that diseases of advanced age due to mutation, including cancer, would be distinctly reduced in frequency. The results of experimental investigations during the last years supply a firm basis for these connections.

All observations point to the fact that the organism "dies daily little by little". As long as cell multiplication surpasses cell death in the individual organs—that means till maturity—the organism grows and with it its functions. After the process of maturation has ended, which can be equated with growth, the number of cells diminish from year to year, and with it also the amount of function of the organs.

The following table shows some typical losses of weight or mass of organs, and the percentage reduction of functions, in comparison of a 75 years old man with a 30 years old:

Weight of brain (normal 1300g)	56 % loss of weight
Water contents	18 % loss of weight
Body weight	12 % loss of weight
Number of kidney glomeruli (or nephrons)	44 % less
Filtration rate of kidneys	31 % less
Blood circulation in brain	20 % less
Heart capacity (at rest)	30 % less
Number of peripheral nerve fibers	27 % less
Number of taste nerves	64 % less
Vital capacity of lungs	44 % less

The organism endeavors to replace and to compensate damages, and the results of wear and tear of its organs. However, the compensatory ability is lost more and more in later years. This represents a chronic progressive process.

Alterations by aging can be shown predominantly in bradytrophic mesenchymal tissues, like cartilage, tendons, blood vessel walls or connective tissues. An important function of the latter is the formation of scleroproteins, the most interesting ones of which are collagen and elastin.

Collagens are water-insoluble, but water absorbing and swelling proteins which are formed in connective tissues, in tendons etc. as well as in the ground-substance of bones; they are resistant against most proteolytic enzymes. All collagens are characterized by a high content of hydroxyproline and only small amount of hydroxylysine. Sulfur-containing amino acids and tryptophan are missing.

In the formation of collagen in the organism out of fibroblast several enzyme systems are involved. Intense investigations during the last years supplied insight into the alteration of the collagen synthesis and of changes of structure during aging.

The synthesis of connective tissue albumen and globulin proteins underly the same controls. The joining together of amino acids to peptide chains takes place over the union of activated amino acids with RNA, whereby the sequence and number is determined by the m-RNA. Under normal conditions the ribosomal synthesis of the collagen-like proteins proceeds under the influence of a certain gen or a group of gens. Collagen is specifically broken down by collagenase to split products.

While we are well informed about the biosynthesis of the collagens, little is known about elastin formation. Elastin is another important scleroprotein which forms the elastic ground substance of the connective tissue. Sequence and number of amino acids are distinctly different from those of collagen. The pancreatic juice contains an enzyme specifically adapted to elastin, elastase (pancreatic peptidase), which splits the elastic fibres into small fibrilli. The enzyme consists of two components which differ in pH optimum. In advanced age the structures of collagen and elastin change. Parallel to it they show also a differing reaction toward the specific enzymes, therefore investigation in this field may be regarded as a model case for biochemistry of aging and for enzyme effects.

Probably the same fibroblast cells synthesize besides collagen also elastin. This would mean that there must exist a double set of ribosomes which also has the corresponding supply of enzymes. Furthermore it may be also possible that certain partial syntheses of collagen and elastin run parallel whereby a set of enzymes remains eliminated. Collagen molecules are not fully stabile, but are influenced in their physical behavior by the number of hydrogen bridges and crosslinks.

In the corium of the aging skin the collagen fibres change into elastin fibrilli. Because the synthesis is taking place in the fibroblasts and is controlled enzymatically, it is presumed that a change in the control mechanism leads to a different selection of the ribosomal peptide synthesis. If one assumes the supposition that all structural proteins are continuously formed during the entire life, the hypothesis is also justified that the qualitative set of enzymes of the cells underlies marked fluctuations during the entire course of life. With this is connected a reduced activity of individual enzymes in advanced age.

Possibly a third protein in the connective tissue is present in old age which differs from collagen and elastin by another combination of amino acids.

To the compounds typical for old age also the "desmosines" are counted, which are formed by synthesis with heterocyclic amino acids and multiply by using the lysis rests of elastin molecules. Some of these crosslinked molecules are resistant against elastase, others show an outspoken lability. In advanced age the affinity of elastin to elastase is reduced. It can be shown biochemically on connective tissue of old men that elastin is being linked to a protein which blocks those reactive groups or centers in the elastin molecule which are sensitive toward the enzyme. This is an example of substrate-blocking by enlargement of the substrate molecule. The elastin of old age has a combination of amino acids different from the elastin of young men (pseudoelastin).

The components of collagen of the skin can be determined by fractioned extraction. The more soluble fractions are alphachains without crosslinks and increase during the middle years of life. They diminish in advanced age in favor of the betachains with cross-links—however poor in H-bridges—which are very resistant to the enzyme collagenase.

With the decomposition of collagen, the fragments are again taken up by the fibroblasts and are resynthesized to collagen, to elastin and similar compounds. In this rotation of the connective tissue components bound to the fibroblast numerous enzymes are taking part upon whose activity and formation in the cell (lysosomes) the primary synthesis of the connective tissue molecules and the secondary resyntheses are dependent.

Investigations are available on the activity of several enzymes during the aging processes. *Barrows* et al. (2, 3) emphasize that the enzyme activity as relative quantity in relation to the number of cells which form the enzymes is of particular importance. They see therefore a relation between the number of cells (measured by the amount of DNA) and of the specific activity. In some cases a dependency of the enzyme activity upon the number of the mitochondria could be determined. The activity of the succinooxidase in the rat kidney diminishes with older animals. *Weinbach* et al. (11) found similar relations for the hydroxybutyrase which loses some activity with increasing age in liver and kidney, whereby also the number of mitochondria is reduced.

It is also known that the aging organism is very much less detoxifiable than the juvenile body. It could be determined by chemotherapeutic tests that the formation of glucuronic acid derivatives as important detoxifying mechanism proceeds slower and more incomplete with advancing age. Either the liver can only incompletely form the corresponding enzymes and in insufficient amounts, or the enzymes do not display their prime activity anymore. This however means that the enzymatically induced enzyme synthesis is failing. The greying of hair is also attributed to a lack of tyrosinase or a loss of activity of the enzyme in advancing age.

The especially important endogenous enzyme—plasmin—also shows an age-dependent influence. Not only the plasminogen synthesis gets diminished during the aging process but also that of the plasminogen activators.

According to *Astrup* (1) about every two hundredth endothelial cell of the intima produces the plasminogen activator, in old organism only every two thousandth anymore. On account of this loss the fibrinolysis gets steadily weaker and with it the hemostatic equilibrium. The sol-gel relationship in the fibrin formation changes in favor of the gel-condition. This results in increased deposits of fibrin upon and into the vessels and in the interstitium of other tissues. While the fibrin deposits in the vessels favor atheroma formation, those in the tissues impair the functional ability of the corresponding organs.

The atheroma formation, as mentioned before, depends on the incorporation of fibrin in the vessels. Subsequently lipoproteins and cholesterols are inserted and thus an organization of the atheroma plaques is initiated. A high level of lipoids in the blood stream hinders the fibrinolysis in vivo and in vitro. Thus also this factor contributes to the formation of arteriosclerotic change of the vessels (plaques). *Hahn* observed that after i.v. injection of heparin the cloudiness of lipemic plasma in vivo was reduced. Later on could be found that in the blood and different other tissues (heart, lungs, fat tissue) "clearing factors" are present which are broken down in the liver. With lipemia the clearing factor is always also demonstrable in the plasma, its activity can be enhanced by heparin. Biochemically, this clearing factor is a lipoprotein-lipase which catalyzes the partial hydrolysis of triglycerides. These are components of all lipoproteins. The enzymatically liberated fatty acids are bound to an albumin fraction and thus transported. The lipoprotein lipases in the tissues break down not only triglycerides derived from the digestive tract but direct also the introduction of lipoproteins and the mobilization of fatty acids into fat depots. The enzyme character of the clearing factor was proven by enzymatic inactivation with the help of another enzyme, the heparinase. Indirect indication of a reduced heparin formation in advanced age can be presumed from the reduction of the number of mast cells in later years.

Besides the frequent appearance of atheromatous changes with increasing age, statistic investigations showed furthermore that the frequency of thromboembolic diseases is also age-dependent. According to *Schumann* venous thrombi of the lower extremities and other locations, pulmonary emboli and, above all, infarcts are in unequivocal correlation with advancing age. *Thies* (10) showed on a section series of 30,493 cases that venous thrombi among 20 year old men are found in only 4%, while with 70 year old men in 24%. Thromboses in other locations of veins show often an abnormal characteristic distribution among the aged. For instance, the thromboses of the vena portae (found in 0.4% in large postmortem statistics) reaches a culmination in the fifth decade, which corresponds regarding frequency with liver cirrhosis in this age group.

Etiological factors for this age-dependent increases of venous thrombosis may be also by the structurally conditioned alterations of the venous walls:

By fluctuation in the circulating time (decrease of the pulse and minute volume).

By the increased viscosity of the blood.
By the increased fibrinogen concentration in the serum.
By the slowing down of the general fibrinolysis.

Because atherosclerosis increases with age, also the frequency of arteriosclerotic thrombi increases. However the frequency of arterial emboli is definitely not dependent on age.

Rheological investigations have shown that the narrowing of the arterial circuit progressing with age, mainly by atheromatous changes with thrombotic deposits, influences unfavorably the relation between pulse and vessel volume. Thereby the "wind kessel" function of the large vessels is reduced or even prevented; as a result of it, the cardiac load is further increased. The measurable slowing down of the blood circulation represents, in connection with the heightened viscosity, an essential factor in the increase of thrombosis tendency.

The frequent fluctuations in the capillary size and their aneurysmatic widenings which can be determined in aging organism further handicap the work of the circulation, increasing cardiac insufficiency and favor thrombi-promoting alterations of wider vessel components.

Among other theories proposed to explain the progressive changes in advancing age are mentioned that certain free radicals produced in the body are causing not only deleterious changes like the chromosome breaks of mutations mentioned, but also the crosslinking of elastic tissue fibres and perhaps other molecules, but also other polymerisations in tissue and plasma responsible for those changes of aging.

The wear-and-tear theory states that all cells at their formation are endowed, according to *Carlson, Pearl* etc. *a priori* with a definite amount of enzymes and potential energy (ribonucleic acid etc.) which during life activities is gradually exhausted in form of kinetic energy. This quantum energy limits functions and life span of every cell. For instance increased warmth and excess food in mice accelerate the "rate of life" of mice or drosophila, while the opposite (*McKay* etc.) retards the aging signs and extends the life span. If the exhausted or dead cells or debris are promptly eliminated or liquified by the circulating or the intracellular proteases, or if their level is raised by an enzyme supply given, the progressing aging process may be delayed.

Alterations of the Plasma Proteins in Advanced Age

About after the fiftieth year of life a reduction of the serum albumen and of the total albumen is found. The globulins in general are not age-dependent, only the gamma globulins are diminished in middle age (about the fifth decade). Physicochemically these alterations may be regarded also as a shift of the plasma proteins toward the coarsely dispersed side. The values of the neuramine acid containing fractions are slightly elevated. Immune electrophoretically it can be shown that fibrinogen, haptoglobulin, beta-lipoproteids, acidic alpha-glucoproteins and the C-reactive proteins are elevated.

Quantitative determinations of the immunoglobulins show a reduction of the IgG and the IgA fractions. The activity of the antibodies toward toxins, bacteria and viruses is diminished, autoinmune bodies (e.g. rheuma factors) increase with age.

That aging processes are also based on autoimmune processes is nowadays a generalized and also justified opinion (e.g. *Walford, Burnett* (5)). Though somatic mutations at first do not cause direct damages, by mutated cells autoimmune reactions are incited which can very easily cause profound effects upon important functions of the entire organism (9). Not only mutation-determined and degenerative alterations of cells, structural changes of collagen and fibrin as well as atheromatous deposits can cause such immune reactions. The reaction products of such immunological processes themselves can again lay the foundation of further degenerative processes, so that immunological chain reactions may ensue. A central position in this process occupy cellular antibodies, especially the small activated lymphocytes.

Their activity manifests itself in an eagerness to attack degenerated or mutated cells. The lymphocytes attack cells sensed as "foreign bodies" by inducing an agglutination of their cellular membrane with the attacked cells. The lysosomal enzymes of the lymphocytes penetrate into the "foreign" cells and destroy them lytically. By this immunological process degenerated or mutated cells, even some cancer cells, are destroyed or diminished in amounts under normal physiological condition. But if such foreign cells are formed in excess so that the cellular immune defense system is not able anymore to fully restore and accomplish its job, finally an exhaustion of the cellular immune defense takes place. This leads to the result that degenerative processes develop which interfere with the function of the affected organ. But the exhaustion of this defense includes also a reduced immunological protection against external influences, a lowered protection against the formation of malignant tumors and a whole row of other pathological conditions. Very recently investigations became known according to which it seems possible by therapeutic measures to reactivate again the exhausted cellular immune defenses. Such stimulation can be accomplished by giving certain polyanions, e.g. heparin and, above all, by megadoses of vitamin A. *Bollag* et al. (4, 4a) could show that the vitamin A application shortens the sequestration period of skin transplantation which undoubtedly is referable to activation of cellular antibodies. *Dresser* could demonstrate also an immune stimulation by vitamin A (7). Also the observation of *Hoefer-Janker* and *Scheef* (8) that the application of megadosis of emulsified vitamin A increases the reaction after BCG vaccination, points in the same direction. A very weak reaction to this vaccination shows a weakness of the cellular immune system.

From all the above statements it is very clear that enzymes, especially proteolytic enzymes, are playing an outstanding part in the aging processes and in their therapeutic countermeasures. Since the mechanism of the entire enzymatic system under physiological conditions is closely interrelated and functions smoothly without friction, disturbance or deviations of the enzymatically governed chain of reactions are bound to put out of equilibrium the function of the cell, of the particular organ and finally of the total organism. Even a pathologically altered biochemical equilibrium is depending upon the functional close interplay of numerous enzymes and their substrates.

From all these facts therapeutic consequences may also be derived. For, when, on the one hand, the endogenous cellular enzymatic activity declines more and more in advancing age, there exists, on the other hand, a strong possibility of substitutive activation by supplying appropriate therapeutic measures. For this purpose proteolytic enzymes are ideal, just made to order, so to say. They are physiological substances, without any side effects, and can be taken by the patient in advanced age, on their own responsibility without medical supervision, over long periods, without any danger of complication, prophylactically as well as therapeutically. Proteolytic enzymes, particularly also in combination with alpha tocopherol acetate, raise, among other beneficial effects, the fibrinolytic activity which is sagging in old age, but is absolutely vitally important, up to a desirable niveau. Thereby a long row of age-determined disease states are prevented. Clinical observations on several hundred "senior citizens" under such protease therapy during the last four years are very encouraging and justify this presumption.

There is no doubt that very comprehensive investigations will still be necessary on large numbers of aging and old people under years of observation to throw light into the many unsettled problems of the aging body and on geriatric problems, but the knowledge acquired by past investigations should already be utilized. This point of view covers the efforts of this chapter.

Lately we could show that by giving larger doses of our proteolytic enzyme mixture all serum lipoids including cholesterol are promptly reduced to normal. The significance of this observation regarding atherosclerosis and its effect on aging of organs like heart, arteries, brain, kidney, eyes, ears, etc. is being investigated at present.

Literature References:

Barrows, C. H., 5. Internat. Congress for Gerontology, San Francisco, 1960

Birren, J. E., et al. The process of aging in the nervous system. Thomas, Springfield, Ill., 1959; Comfort, A Biology of Senescence, Rinehart, N.Y., 1956

Montagne, W., et al. Biology of Hair Growth, Ac. Press N.Y. 1958

Stieglitz, E. J. Geriatric Medicine, Pitman & Sons, London 1954; A.A.G. S. 65, 1960 "Aging"

1. *ASTRUP*, T., Biology of Plasmin, Schattauer, Stuttgart, 1970
2. *BARROWS*, C. H., et al., J. Gerontology, *13*, 351 (1958)
3. *BARROWS*, C. H., et al., J. Gerontology, *15*, 130 (1960)
4. *BOLLAG*, W., et al., Schweiz. med. Wschr., *101*, 11 (1971)
4a. *BOLLAG*, W., et al., Schweiz. med. Wschr., *101*, 17 (1971)

5. *BURNETT*, F. M., Lancet, *7668*, 358 (1970)
6. *HOEFER-JANKER*, H., Aerztl. Prax., *23*, 2805 (1971)
7. *DRESSER*, D. W., Nature, *217*, 527 (1968)
8. *HOEFER-JANKER*, H., Aerztl. Prax, **21,** 5238 (1969)
9. *PLIESS*, G., Actuell. Geront., *1*, 3 (1971)
10. *THIES*, H. A., Zbl. Phlebol., *3*, 146 (1964)
11. *WEINBACH*, E. C., et al., Gerontology, 3, 253, (1959)

VIRUS-DISEASES AND ENZYME THERAPY

While we are able nowadays to combat bacterial infections of men and animals to a great extent by chemotherapeutica, as to a causative therapy of viral infections we are still in the beginning.

Though by prophylactic measures like active or passive immunization, some dreaded epidemic virus infections, like smallpox or poliomyelitis, have been brought under control in most civilized countries, many virus infections, even those of a trivial nature, can at present only be treated symptomatically.

The relatively young virology is now in a period of speedy development on account of the technical progress. Mainly the use of refined electron-microscopic methods and the advances in cell- and tissue cultures led in the last few years to a steady increase of our knowledge. Smaller and smaller virus particles become visible, new culture methods for growing them were developed and the functions and effects of them studied. Lately information appeared that new particles were discovered whose size is still below the smallest known viruses. This would offer new aspects for the etiology of several diseases of unknown etiology.

A possible control of viral infection cannot be thought of without some knowledge of the biology, chemistry and structure of the virus particle. Fundamentally, the structural parts of all viruses are similar. Their forms vary between stablike, balllike, polyhedral and, in case of bacteriophages, "spermlike". Also their size varies considerably. According to our present state of knowledge, a virus is a lifeless chemical unit whose structure and chemical constituents resemble, in a primitive simplification, those of living cells. Fundamentally there are four characteristics which are common to all viruses, whether pathogenic to animals or to plants or saprophytes.

1.) they are all of submicroscopic size,
2.) they have no known metabolism,

3.) they contain nucleic acid,
4.) they can only multiply within a living cell.

On the example of bacteriophages, the activity, mechanism of multiplication, the so-called identical reproduction, and other functions can be best studied. The impressive progress of virology and beyond this in the molecular biology of all living cells during the last decade has been much helped by experiments with bacteriophages, besides the study of the tobacco mosaic virus and some plant tumors.

Construction and Parts of a Virus

A virus represents a giant molecule which chemically consists of nucleic acid and usually proteins, combined, in larger viruses, with lipoids or carbohydrates. The viruses possess enzymes which govern all processes during multiplication and cell infection, while the nucleic acid directs them and all their activities. Among the large viruses of the psittacose-lymphogranuloma groups the minutest structure of the virus particles approaches almost that of a living cell, there even exists some sort of a cell membrane.

The nucleic acid is apparently the main viral substance. It forwards the genetic information, it controls the forces which the host cell induces to form new specific viruses identical to the original intruder. In the virus itself the nucleic acid is so formed or protected that it cannot be attacked by nucleases. There are virus types which consist only of nucleic acid.

Nucleic acid is composed of a large number of nucleotides which in the simplest types are joined by phosphoric acid, a pentose and a base. The pentose can be present as ribose or desoxyribose.

Virus proteins are composed of polypeptide chains. The nucleic acid as material carrier of the biochemical and morphological information is surrounded by the covering of a protein film. The protein covering plays an important part in attraction and attachment to the host cell and the penetration of the virus nucleic acid into the cell. Otherwise it functions as point of attack (antigen) for neutralizing antibodies.

The Multiplication of Viruses

Exactly how the infection of the host cell by the virus is taking place can not as yet be clarified. With bacteriophages we already have exact information and it is to be presumed that the scheme with other cell viruses is essentially similar or related in a varied form. At first the virus attaches itself to the host cell and begins to penetrate it with the help of its mucinolytic enzyme. From this moment on the attacked cell is unresponsive (protected) to further virus infections, an interference exists. Now for a certain time no virus particle can be identified in the infected cell. This latent period is different with different viruses and is called eclipse.

Now in the nuclei of the host cell the reproduction of the parts of the virus nucleic acid is taking place, whereby the first intruded virus substance functions as a matrix. The protein part of the virus, the so-called second virus substance, is produced at the same time in the protoplasma of the host cell. This is followed by isoformation, by which the virus nucleic acid covers itself with its protein film. Now the road is free for the birth of the newly formed virus particles which, after disintegration of the used up host cell, enter the serum or plasma, ready for new infection. In many cases, however, the host cell does not perish; it survives the infection.

Viruses therefore do not multiply by simple division; the infected cell completely neglects its own activities regarding substance synthesis and metabolism, it exhausts all its forces with the requirements of its infective agent. The particular reason why a cell is suddenly forced to reproduce virus substances instead of its own constituents is unknown. But it is presumed that the information of the virus nucleic acid consists of the building of a new informative system between virus and host cell.

Epidemiology and Size of Viruses

Viruses are found about everywhere on earth, but only a very small number of types are pathogenic, most of them are apparently saprophytes. The three virus types causing poliomyelitis belong to a group of over 50 different entero-viruses; those causing colds, the different adenoviruses, Coxsackie and Echo viruses; the viruses causing the different infections of: hepatitis, encephalitis, herpes simplex, herpes zoster, yellow fever, chickenpox, German measles, and the other exanthemata, and of dengue fever etc. belong to the most important human virus infections.

Also many zoonoses, like psittacosis, ornithosis, rabies, hoof-and-mouth disease etc. are caused by viruses which can be transferred from animal to man.

In general, virus diseases play a very prominent part among the animal diseases. Every year they cause substantial economic losses, often in form of epidemics. The hog cholera, infections like leucosis, Newcastle disease, Mareks chicken paralysis, laryngo-tracheitis, erythroblastosis etc. as well as broncho-pneumonia of horses, calves and pigs, also dog diseases (distemper, pneumonitis, hepatitis, laryngoenteritis of cats, myxomatosis of rabbits. No animal (fish, insects etc.) is probably immune against some virus infection.

In different experimental animals several types of malignancies can be produced which are caused by viruses. The polyoma virus, for instance, causes in the same animal carcinoma and sarcoma. In hamsters, the monkey virus SV 40 causes fibrosarcoma and carcinoma on different places, fowl viruses can infect rodents, producing cancer etc.

The modern virology was developed from study of plant viruses by *Iwanowski* in 1892. This scientist demonstrated that the juice of plants infected by the tobacco mosaic disease can infect other plants. Later it was found that plants can be infected by the juice filtered through porcelain, it was cell free. Also other plant diseases are caused by various virus infections; the bacteriophages infect their corresponding bacteria.

Regarding the size of viruses, some are on the borderline of light microscope-visibility (vaccinia virus, myxoma), they have the size of 200-300 milli microns.   Viruses of colds, herpes or influenza, also dog pneumonia etc. have the size of 100 mµ-160 mµ, while polio, Echo, yellow fever viruses are about 15 mµ long (size of a staphylococcus is 1000 mµ).

Virus and Cancer

All living cells of men, animals and plants divide first in a certain rhythm. This continues until the available space is filled out. Then cell division stops. This means that a cell is able to discontinue further growth at a three-dimensional contact inhibiting the mitosis apparatus. This phenomenon of every normal cell in an organism is regulated by some feed-back enzyme system, the so-called contact inhibitor factor. In tissue culture this can easily be observed. But if in a tissue culture a neoplastic mutation appears, induced, e.g. by virus infections, the contact inhibition factor loses its effect, the inhibition on the cell surface fails, the mutated cells multiply without limits and assume the main characteristics of malignant tumor cells, the invasive growth. This tumor character is inheritable to succeeding generations. In contrast to chemical or physical cancerogenic influences which are brought about by damage to certain genes, the mutation by viruses is probably caused by the fact that into the nucleic acid molecule a part of a foreign genome has been included. This can be demonstrated immune-biologically.

If into a grown-up animal special oncogenic viruses are injected and, later on, tumor cells implanted which were caused by the same virus, the implantation cells do not grow and no tumor starts forming. But if the same tumor cells are injected into an animal not previously treated, a tumor is developing.

In a newly born animal the immune apparatus is not yet fully developed, although with the first mother's milk, the colostrum, a certain amount of specific immune bodies is transferred, as far as the mother animal was able to develop them. The concentration of them and with it the chances of resistance are, however, rapidly lost. Finally the contact of the young organism with corresponding antigen leads to the formation of a new defense system of its own body. Therefore, if to very young animals tumor cells are given, in most cases tumors arise, though they may often appear only after months or years. This long time of latency may have different possible explanations.

The lysogeny (see below) which some viruses have, may play a part. It could also be that viruses lie dormant till, through certain irritations, cancerogenic insults or changes due to age, some cells begin their excessive growth. Furthermore the tumor growth may at first have taken place so slowly that it could not be detected. Also a highly active immune apparatus could destroy all steadily new arising tumor cells by its antibodies. Only after the antibody formation in the aging or exhausted organism is getting out of step with the new supply of tumor cells, it does not catch up any more with the production of antibodies and finally a tumor growth has its way.

The adherents of the virus etiology of certain human leukemias and other malignomas rely on such observations of animal malignancies caused by viruses. But no definite virus could be ascertained electron-microscopically so far with human malignancies.

The phenomenon of lysogeny mentioned before is a fact which the following example may explain. If virus particles are brought into a host cell, sometimes it does not bring about the start of the expected infection. In the cells themselves the virus substance cannot be identified, neither histologically, serologically nor chemically. In reality, however, virus substance is indeed somehow "inherited" from cell to daughter cell. Predisposing moments (stress situations) may cause a sudden outbreak of infection. Frequently such infections run a most stormy, almost explosive course.

In a virus disease (hog influenza) lysogeny could sometimes be involved in complicated ways. The virus of this influenza is also taken up by lung worms which by chance are parasites in acutely sick pigs. The lungworms lay "lysogen infected" eggs which are eliminated through the digestive tract and eventually taken up by earth-worms. In this worm the development cycle of the lung worm larvae takes place. If such earth-worms are again taken up by a pig, lung worm larvae find their way finally into the pig's lung. In case of predisposing moments the virus influenza breaks out most violently and suddenly like with a bang. During the entire development cycle there is no specific virus substance in any form to be found. The now active virus was all the time so-to-say "masked". Whether the lysogeny, however, is applicable to the pathogenesis of cancer, needs clarification.

Therapy of Virus Diseases

In the recovery from infectious diseases antigens and antibodies play sometimes an important part. We know that after a number of different virus diseases an absolute and lifelong immunity remains (e.g. measles), while bacterial infections are mostly followed by only relative immunity which after a time often finally disappears. Also the success in using specific vaccines as neutralizing antibodies prove that the antigen-antibody principle holds true also with viruses. The virus-infected organism, however, produce antibodies only at a relatively late period of the disease. They can therefore interfere only very late with the course of the disease. The virus itself is by that time, however, already within the body cells, protected against the action of at least the humoral antibodies. This explains the known great therapeutic difficulties of these diseases, for viruses are obligatory intracellular parasites.

Isaacs in London observed that in cultures of virus-infected cells these partly perish, but after some time another group of the cells is not destroyed anymore but recovers or remains normal. He found that those cells under the influence of the infection produced an antiviral substance which he called Interferon. This substance is produced by cells of most animals and men, it is the product of an unspecific reaction of the virus-infected cells. The action of the interferon consists of the production of a temporary resistance against the infection of the neighboring cells. By these means the further spread of the infection is stopped, which presumably leads to a cure. Every animal produces its specific interferon.

Interferon suppresses oxydative processes inside the cell, and this blocks the liberation of energy necessary for viral synthesis. Cortisone or oxygen supply influence the efficiency of the interferon in a negative direction, an increase of temperature produces higher efficiency.

Possibilities for inhibition of virus multiplication are given theoretically either extracellular in the blood stream, respectively in the blood itself, in the stage of attachment or during the biosynthetic process within the cell after the process of the virus eclipse. Theoretically and in vitro it is possible to block the virus synthesis in the cell by the intake of nucleic acid-antimetabolites. But every substance which suppressed the synthesis of nucleic acid in the cell and with it also the formation of virus-specific substances, damages at the same time the host cell, because all building blocks of the virus and the energy for virus formation originate in the cell itself. The constituents of the host cell and of the virus show a close homogeneity.

Specific efficient chemotherapeutica against viruses do not exist at present. Innumerable substances and remedies were tested in this respect. In the veterinary medicine some had moderate success, but a genuine active principle against virus infection could not be produced. Also the interference with the intracellular identical reproduction by means of interferon promises only little success.

Bieling and *Gsell* (3) assert that principally it must be possible to dissolve the protein covering of the virus and thereby prevent the attachment to the host cell and the specific chemotaxis. By such measure the subsequent infection would be prevented.

Also *Weidel* (27) writes that the influencing of the nucleic acid molecules hardly represents a point of attack for therapy; that it would make more sense to work on the protein enclosure.

It follows from these considerations that the destruction, lysis or inactivation of the accessory component of the virus particle, its protein film with its enzymatic principle, would be a real success-promising way to conquer the virus diseases. The logical solution of the problem of an efficient therapy and a dependable prophylaxis is the raising of the proteolytic potential in the blood and the intercellular plasma. Till now the real success with several virus infections encourage a continuation on this road.

Proteolysis is only possible without side effects in the organism by means of raise of proteolytic enzymes. The elevation of the proteolytic niveau in the blood and plasma represents therefore an efficient means to control virus infections.

Indeed *Wild* and *Brown* (28) found that hoof-and-mouth disease virus inactivated by trypsin, completely loses its infectivity, although the ribonucleic acid remained intact.

Cleeland (4) found similar trypsin effects with six different influenza type A viruses; polio and other entero-viruses could also be inhibited by enzymes—in this case nucleases.

After applying digestive enzymes to varicella viruses, *Bieling* and *Gsell* (3) saw the inner structure of the virus getting more distinct, a sign that the protein film was dissolved or at least profoundly changed.

Wolf and *Benitez* (29) accomplished similar results with polio and vaccinia viruses by using the enzyme mixture Wobe-Mugos.

Shields (25) used chymotrypsin orally against herpes zoster and accomplished a very prompt freedom of complaints.

Also *Rosanova* (18) reports about success with herpes zoster, virus pneumonia and grippal infection after using the proteolytic enzymes.

Meiers (14) advises such enzymes also for local application in herpes zoster; about clinical use of lysozyme in a measles epidemic writes *Christofani* (6) who determined a shortened duration of the exanthema and shorter fever time.

Other investigators (7a) found by lysozyme therapy of spinal polio improvements of spontaneous movements, muscle tonus and reflexes.

Götz (10) used for herpes simplex, besides vaccination therapy on recidive places, locally the enzyme salve and saw in several cases the cessation of recurrent rash or at least prolongation of intervals. Impressive results were also seen in parotitis cases (mumps) in which a prompt improvement took place if the enzyme mixture was given in time.

Miller et al. (15) had already in 1956 successfully treated mumps with or without complication of epididymo-orchitis with streptokinase. Investigations over a long time led to the conviction that colds caused by virus infection could be successfully reduced if during the cold season prophylactic proteolytic enzyme mixtures are systematically taken.

It is known that proteases, especially trypsin, are able to dissolve almost all native proteins as long as they are not components of living cells. Living normal cells are protected against lysis by an inhibitor mechanism.

Virus particles are cell parasites which in their extra-cellular phase do not show any of the characteristics of life. Therefore they cannot in this stage develop any protective inhibitors against proteolytic enzymes. We may conclude that the protein cover of the viruses during their extracellular phase can by proteases be dissolved or at least inactivated. This is bound to result in a loss of infectivity of the virus.

It is possible that the loss of infectivity is also based upon another effect. *Schmähl* (20), *Ruhenstroth-Bauer* (19) and *Coman* (5) report that cancer cells possess a "stickiness". This is due to the fact that fibrin deposits are precipitated on the cancer cell membrane. This stickiness enables it to attach itself to the vessel endothelium and later to start its invasive or metastatic growth. If this fibrin film is eliminated, e.g. by a proteolytic (fibrinolytic) active principle, the adhesion of the malignant cell and thus the formation of a tumor could not take place.

This mechanism could also apply to viruses. If these possess a stickiness, irrespective whether it is caused by fibrin deposits of the host cells or originated from the protein film of the virus particle, the loss of stickiness would also lead to a loss of ability to adhere. This process would deprive the virus particle of a chance of cell invasion, a beginning or already existing infection would be stopped, so that the bodys defense powers could easily get rid of the "impotent" virus particles.

These theoretical considerations induced *Wolf, Ransberger* and *Benitez* (17, 29) to thoroughly investigate antiviral effects of the enzyme combination Wobe-Mugos, a mixture of several proteases and activators. At the Biological Research Institute of New York experiments were set up to try to demonstrate the lysis of the protein film covering different pathogenic viruses. On account of technical difficulties this was not successful. Then this problem was taken up with the tobacco mosaic virus, according to the following technique: the enzyme mixtures were dissolved in isotonic salt solution and injected into the stem of tobacco and bean plants. Above the spot of injection a plastic globe or bell covered the plant and was tightly attached to the stem. A slight vacuum was produced within the globe or bell in order to intensify the resorption flow from the stem into the leaves. After the enzyme infusions for ½ hour, the leaves were infected with tobacco mosaic virus two hours later.

Sixty percent of the plant leaves did not show any visible infection, about 40% had pigment changes in a small area around the injection point. These, however, remained stationary, while all control plants showed the signs of infection on large areas of their leaves.

In another series of experiments, tobacco plants were first infected till distinct signs of the infection were present. Three to fourteen days later the enzyme solution was injected. This caused a prompt stop of further signs of the infection, and the spots lost their infectivity.

In cell cultures of the C-57-mouse the vaccinia virus causes easily detectable agglutination reactions. If to the culture medium shortly before the infection with the virus 25-30 gamma of the enzyme mixture is added, an agglutination either does not take place or only to a slight extent, and there is no necrotization in the cell culture.

With polio viruses or He-La-cells there is a similar agglutination reaction. But also in these cultures no agglutination and no necrotization takes place if to the culture 25 gamma of the enzyme mixture is added. Similar results are seen in using human amnion cells.

These and similar other experiments have only one explanation and the results are easily reproducible. *Bayerle* (2) tested the effect of the enzyme mixture upon the Ryley virus of the mouse. The intramuscular injection of the protease mixture caused a drop in the LDH increase which normally is pathognostic for the take of infection. *Glock* (9) investigated the therapeutic value of the enzyme preparations in a viral infection of a large number of beef cattle, an epidemic virus pneumonia with complicating secondary infections which frequently ended lethally. High doses of sulfonamide or antibiotica, remedies for the cough and circulation brought about only insignificant improvements of the prognosis. Parenteral therapy with the enzyme mixture combined with antibiotics resulted in a complete restoration of health within two to five days among the 240 animals.

Similarly definite and outspoken were the results of the therapy of pneumonia in hogs (*Friedl* (8)), the virus bronchitis of horses and young pigs (*Glock* (9).

Most impressive was an extensive series of experiments performed by *Dunkel* (7) who in charge of the Veterinary Medical Development Program in Chad (Africa) supplied the enzyme mixture mixed in the dry feed in huge fowl colonies where chicken leukosis was rampant. While practically all treated animals recovered, all the controls were killed by the disease. Further investigation in such epidemics are very desirable to clarify this effect.

Herpes Zoster

Also in human therapy impressive results could be accomplished in a series of virus infections. Prominent among them is the successful therapy of herpes zoster with proteases. The clinical symptoms of fresh infections as a rule are reduced or eliminated within two to four days after starting the use of the enzyme mixture. Cases which had the infection for a longer time sometimes required treatment of at least six to ten days. The neuralgias which often develop at the same time or following the rash disappear very soon or do not become manifest. The skin efflorescences heal within a few days, secondary infections do not take place (10).

The average doses of Wobe-Mugos in this disease consisted of daily two suppositories, two times four tablets, and one ampule intramuscular. Also the enzyme salve applied to the rash is advantageous. Lately the use of clysmas of 3 g Wobe-Mugos shows prompt cures (retained in rectum).

In especially severe forms and in refractory cases of shingles, the possibility of a basic disease which lowers the body's resistance, like lymphogranulomatosis, leukemia or carcinoma, should be investigated by differential diagnosis, because in a large number of such cases later on a malignancy may be found. Reports about this fact were published in 1955 by *Wyburn* and *Mason* and in 1964 by *Moynihan*. Similar reports were published more recently by *Rawls* et al., who found a connection between zoster and cervix carcinoma. Statistics also show that herpes zoster is found much more frequently in severer form among patients suffering from Hodgkins disease.

Herpes Simplex

Similar results with the enzyme therapy like in herpes zoster are reported in herpes simplex. *Götz* (University Clinic Münster) (10) writes about the topical application of the enzyme preparation in recurring herpes fever blisters. He and others are using the local application of Wobenzym in salve and find a significant prolongation of intervals, in some cases a complete cessation of efflorescences. The treatment of herpes simplex cases consists of daily use of one suppository and 4 tablets twice a day.

Warts

The virus etiology of most juvenile warts has been definitely established, but the mode of infection, any reservoir of the virus in the body, immunity reactions, latent periods etc. are still largely unsolved problems. It was also determined that viruses can be present or absent in the warts. Certainly recently formed warts contain much more virus particles than those which persisted over a year. These facts explain the often observed resistance against therapy of long existing or recurring warts.

For the therapy of fresh warts often the local application of the enzyme salve is sufficient. In cases of massive dissemination of warts daily four times two tablets should be given in addition, also the direct injection into the wart promises success. In this case the greater part of the dissolved contents of an ampoule is injected two to three mm deep through the wart (1a).

Bronchopneumonia and infant pneumonitis due to virus infections, influenza and colds, epidemic parotitis, rubeola, morbilli and others are distinctly favorably influenced by this enzyme therapy. The course of the disease is usually considerable shortened, the symptoms are milder and extensive complications are avoided. Mixed or secondary infections caused by bacteria require often corresponding antibiotic treatment besides.

In affection of the respiratory tract, tenacious bronchial and nasal secretions are liquefied by proteases and then easier coughed up. Also the antiphlogistic effect of proteases help the healing processes. For the therapy of affections of the lungs and bronchial tract, the following daily schedule is advised depending on individual cases:

> two enzyme clysmas
> two times one suppository,
> up to 10 protease candies,
> one ampoule intramuscular.

The doses for infants and children is corresponding less. In diseases of the upper respiratory tract, instead of the tablets up to 10 candies daily are advised. A sufficiently large dosage, especially high initial dosage, often clinches the success.

In the virus infections mentioned, like in all febrile states, the supply of the epitheliotrope Vitamin A from the liver is blocked. Therefore it is always desirable to give also Vitamin A in sufficient amounts.

Influenza and the different forms of general colds are almost always mixed infections. The primary virus infection being joined by a secondary infection of bacteria cannot be cured etiologically by enzymes, for live germs like all other healthy cells are naturally protected against proteases, although the results of bacterial processes, like catarrhal exudates, necroses etc. can be liquefied and removed by them. Experiments on a broad basis in England showed the result that with patients, man or animals, only after temporal local or constitutional lowering of resistance first, a virus infection was found before the invasion of the ubiquitous germs could enter the epithelial lining. The viruses, so to say, paved the path for the secondary bacterial invasions. These experiments covered the disease conditions which are usually referred to grippe and cold viruses. Experiments showed that colds, which always represent mixed virus-germ infections, could be successfully treated or prevented by increasing the proteolytic enzyme potential of the plasma.

This, however, has to be done promptly, probably within the first 24 hours. Since the start of the infection can be exactly established only in the rarest cases, as it has the symptomatology of any allergic reactions, a continued prophylactic increase of the proteolytic enzyme niveau offers the best chances to avoid them.

The Biological Research Institute of New York started statistical observations regarding the prophylactic value of proteases against colds, which gives some interesting practical information, though the number of persons in the experiment is not sufficient for statistical purposes (150 experimental persons and the same number of controls). Of 90 persons who suffered the last few winters from severe colds or grippe, 68 were free of any colds while the course of the colds with the other 22 patients was mostly very slight. All of the experimental persons had taken, during the cold seasons in the last two years, a prophylactic dose of four to eight enteric coated tablets daily of the enzyme mixture. Of 60 patients who took at the first typical cold signs twice the dose over three to four days, only eight suffered the usual course of colds. The control group had the usual number of severity (85%) of colds and grippe infections like in former years.

There may be two reasons to explain the beneficial use of an elevation of the proteolytic level of the plasma by taking proteases as prophylaxis or as therapy of all respiratory infections. First its antiviral effect, mentioned above, and secondly its antiinflammatory effect. Regarding the latter, it is generally accepted and experimentally demonstrated that a healthy mucous membrane is protected against most bacterial invasions, but an irritated or inflammatory state of it opens the doors for invasion of pathogenic germs, which is closed again by antiinflammatory enzymes.

This explains also our observations that infection-prone general conditions, like mongolism, cretinism, diabetes or all "rundown" cases show so much better protection against germs, or their germ infections are cured so much faster, if they take the enzyme therapy straight along. The enzymes deprive the bacteria of their hiding and breeding place. Such small enzyme doses raise slightly the proteolytic niveau and do not elicit inhibitor formation.

Certainly very extensive trials on large population groups are necessary to determine statistically the exact time and dosage for an efficient enzyme prophylaxis and therapy of virus infection, pure or mixed; but the clinical results so far suggest the way such double-blind investigation should be tried.

The surface seems hardly to have been scratched as yet on the potentialities of proteolytic therapy of diseases, definitely or possibly caused by virus infections, in particular, and all other diseases and physical handicaps in general, which may be helped by proteolytic enzyme preparations. The future field of this new physiological remedy seems almost limitless and many experienced students of proteolytic enzymology regard it as "The Future of Practical Medicine".

Literature

1. *AMBROSE*, E. J., Nature, *177*, 576 (1956)
1a. *BARRI, ANA LOPEZ*, Act. Dermatol., 7, 91 (1967)
2 *BAYERLE*, H., Personal Information (1966, 1968)
3. *BIELING*, R., Die Viruskrankheiten, Menschen
   GSELL, O., Barth-Verlag, Leipsig (1964)
4. *CLEELAND*, R., Proc. Soc. Exper. Biol. Med., *112*, 913 (1963)
5. *COMAN*, D. R., Cancer Res., *21*, 1436 (1961)
6. *CHRISTOFANI*, M., Minerva med., *58*, 2282 (1967)
7. *DUNKEL*, R., Personal Information (1968)

7a. FERLAZZO, A., LOMBARDO, G., TIRALOSI, G., Mal. Ifer., 7, 131 (1961)
8. FRIEDL, L. W., Personal Information (1968)
9. GLOCK, H., Personal Information (1967, 1968)
10. GÖTZ, H., Münchn. med. Wschr., 22, 1240 (1967)
11. HEINE, K. M., Münchn. med. Wschr., 107, 1038 (1965)
12. HEINESSEN, W., Behringwerk Information, 37, 53 (1959)
13. ISAACS, A., LINDEMANN, J., Proc. Roy. Soc. B, 147, 259 (1957)
14. MEIERS, H. G., Med. Welt., 19, 1973 (1968)
15. MILLER, J. M., et al., Mil. Med., 118, 31 (1956)
16. PREISIG, R., BONIFAS, V., COTTIER, H., Schweiz. Med. Wschr., 97, 1373 (1967)
17. RANSBERGER, K., Lecture at the Sigma X Conference, Colorado State Univ., 24, 2 (1967)
18. ROSANOVA, A., Unpublished (1967)
19. RUHENSTROTH-BAUER, G., Klin. Wschr., 44, 39 (1966)
20. SCHMAHL, D., Med. Welt., 11, 544 (1964)
21. SCHONEBERGER, M., Behringwerk-Information, 37, 42 (1959)
22. SCHONEBERGER, M., Behringwerk-Information, 37, 109 (1960)
23. SCHRAMM, G., Behringwerk-Information, 25, 157 (1952)
24. SCHRAMM, G., Die Biochemie der Viren. Springer-Verlag, Berlin-Göttingen-Heidelberg (1954)
25. SHIELDS, T. L., Current News in Dermatology, Aug. 1962
26. TYRELL, D. A. M., BYNOE, M. L., HITCHCOCK, G., PEREIRA, H. G., ANDREWS, D. H., PARSONS, R., Brit. med. J., 1, 606 (1968)
27. WEIDEL, W., Virus and Molecularbiologie, Springer-Verlag, Berlin-Göott:ngen-Heidelberg (1963)
28. WILD, T. F., BROWN, F., J. gen. Virol., 1, 247 (1967)
29. WOLF, M., BENITEZ, M. H., Personal Information (1965)

ENZYME THERAPY OF CANCER

Proteolytic enzymes in the treatment of malignant tumors have been used in historical times already. Long before the discovery of America by Columbus; medicine men of the Indians applied fruits and leaves of the papaya plant to malignant tumors, they used local enzyme therapy empirically. It was known that fresh papaya fruits favorably influenced inflammations and edemas, that wounds, burns, bruises or infections healed faster and pains subsided sooner, also that malignant tumors responded sometimes to this therapy.

About the year 1820 *Physick* in Philadelphia was the first to use proteolytic enzymes in the form of stomach juice for surface cancer with good results. In 1836 *Schwann* isolated pepsin from stomach juice, in 1871 *Purden* and in 1888 *Douglass* applied the enzyme pepsin to ulcerated cancerous lesions. At the end of the 19th century the first attempts were made to give pepsin intramuscular and trypsin intravenous. In the year 1902 the enzyme therapy of cancer received a decided impulse when *John Beard* began cancer treatment with enzyme extracts of the pancreas. His therapeutic successes caused great excitement. By 1906 he had used trypsin, amylopsin and other unknown enzymes of the pancreas in treating different cancers. His comprehensive book: *The Enzyme Treatment of Cancer* stirred up great interest among many scientists and clinicians who soon went about to develop further this therapy and to use it extensively.

Beard, the leading embryologist of his time, concluded from his studies for many years of the embryonal development of animals that during the progressive differentiation of the developing organs undifferentiated polyvalent "sex" cells wander from the trophoblast mainly through the mesoderm within the embryo to their goal of destination, the gonads. Countless cells of them get stuck on this voyage between somatic cell aggregates. These everywhere dispersed embryonal trophoblast isles remain dormant, according to *Beard,* and do not multiply during the entire life span of the individual. However, the one or the other cell can by specific irritants (cancerogens) start a cell division and thus form a functionless tissue island, a tumor. Indeed, American scientists, like *Hayflick,* during recent years were able to identify in tissue cultures of the different organs such scattered cells, about 0.25 to 0.5% of the total, the least number in heart tissues. They differ morphologically from the other cells by absorbing more dyestuff, but without mutagenic or cancerogenic irritations they remain dormant. When their mitosis starts, they stain darker than the somatic cells, also typical characteristics of tumor cells appear. Compared with normal cells, they are potentially immortal, i.e. they multiply without any restraint through hundreds of subcultures while all somatic cells in the cultures "age", they lose their ability to further divide and die after not more than 50 mitoses.

Soon physicians all over the world were interested in the theoretical as well as the excellent therapeutic results. The preparations used consisted mainly of freshly prepared pancreatic extracts (*Campbell, Goeth, Duprey, Curtfield, Marsden, Meggit, Cleaves, Shaw-McKenzie, Little, Bainbridge*).

Hald, Pusey and *Blumenthal* reported before that intra-tumoral injections of trypsin would bring about a relatively fast softening of the tumor, with aseptic liquefaction.

However besides the therapeutic successes also side effects of a pyrogenic and toxic nature appeared. Finally they began to produce pancreas extract industrially for a longer shelf life of the product. But it was unknown at that time that after a few hours of storing at room temperature the enzyme activity of the liquid extracts was lost. Their use resulted in a deterioration of therapeutic results obtained and led finally to the fact that the enzyme therapy of cancer was forgotten, or rather fell into hibernation.

Only much later the factors were recognized which ruined the confidence in *Beard's* therapeutic development: the instability of the enzyme products, their antigenicity and the impurity of the extracts used that time as well as their contents of pyrogens and toxic admixtures.

When much later it became possible to produce crystalline and pure enzymes, the therapeutic application could be resumed again in larger amounts. *Sumner* crystallized in 1926 Urease, *Northrop* in 1930 Pepsin, *Northrop* and *Kunitz*, Trypsin. Thus it became possible to stabilize the enzymes and to eliminate pyrogens and other toxic substances.

1934 *Freund* in Vienna discovered that in the serum of people or animals free of cancer chemical substances existed which were able to dissolve cancer cells, while the blood of cancer patients was lacking in this ability. Besides that, the *Freund-Kaminer* team also found that the serum and urine of cancer patients not only was lacking in cancerolytic property but that cancer cells were even protected by it against dissolution by normal serum and produce a cancer-protection substance in the serum.

If, for instance, to normal serum half the amount of cancer serum is added, the former loses its proteolytic capacity against cancer cells. *Freund* isolated this water-soluble, thermolabile substance from the serum and urine of men and horses free of cancer; he called it "Normal Substance" and used it with partly good results as parenteral therapy on inoperable cancer cases. Furthermore, this phenomenon led to the development of the *Freund-Kaminer* reaction. This test indicated that the serum of cancer-free people and animals dissolves a large percentage of cancer cells in a fresh cancer suspension (later he used for this test heat-killed necrotic cells) or it changed them markedly. However, cancer serum hardly affected them, it even protected them against disintegration by normal serum. *Kretz* and *Benda* could verify these facts, also *Klein* and *Lustig*. *Freund* had found 30 years earlier that cancer serum possesses these cancer-protective qualities which are derived from abnormal fatty acids found in the intestinal tract. His directions to fight the cancer disease by diet (total elimination of animal fats and fermentative foods and by elimination, as much as possible, of the "abnormal" (acid-fast) colibacteria, which live in the colon of cancer patients, by "intestinal antiseptics" like menthol, or by enemas) are based upon these investigations.

Freund and *Lustig* showed in 1933 on tar cancers of mice that a cancerophile diet hastens the tumor development and the animals died earlier, while a cancerophobe diet protects 50% of the mice. On account of the beginning World War, *Freund* and *Kaminer* were forced to discontinue their activity in Vienna and therefore had no chance anymore to identify chemically the isolated Normal Substance. By *Christiani* in Vienna it was later identified as a cytolytic enzyme and, independently shortly before, by us (*Wolf*) as a proteolytic-lipolytic enzyme.

Christiani worked on the problem of closer examination of the cancerolytic enzymes and could demonstrate that the "normal substance" is in fact a hydrolytic enzyme which he called "solving enzyme". He proved 1938 that this solving enzyme is bound to the albumin fractions and is thermolabile. It is present in the serum and urine of healthy people and animals (horses), but absent in the serum and urine of cancer patients. Later on *Christiani* found in the serum of cancer patients some of the inhibitors of the solving enzyme. They protect cancer cells against the solving enzyme and are produced by the cancer cells. Also *Freund* knew about such inhibitors which were named "protective substances". Cholesterol esters, e.g. cholesterol-butyrate or cholesterol-succinate have this protective action and indeed are identical with those formed by the cancer cells.

Christiani also could show that the inhibitors, identified by him, could themselves be inactivated by a number of substances, like oxydation products of ergosterol or the 7-dehydrocholesterol. Such substances have acidic character. He called them de-activators. In vitro they were able to prevent the attachment of the inhibitor—the protection of the cancer cell—to the solving enzyme, as well as to free again an already blocked enzyme. Further investigations proved that the de-activator present in healthy people is bound to the globulin fraction of the serum. The de-activator is synthetized from 7-dehydrocholesterol by means of the enzyme ergosteroloxydase, which can not be demonstrated in cancer tissues.

Since then numerous scientists were able to prove that the serum of healthy men and animals is rich in proteolytic, lipolytic and amylolytic enzymes. Patients with more or less active inflammations or infections have in general a lower enzyme potential, but by far the lowest enzyme content as a rule is found in cancer cases (32, 33). Since in precancerous and in earliest stages of beginning cancers the enzyme niveau in the serum appears very reduced, it seems most probable that low enzyme values represent a predisposition or condition for the malignant process. Sometime the low level is inherited, in some cases it may be caused by chronic infections, damage to pancreas or other diseases, possibly also by faulty nutrition.

Gaschler et al. determined the proteolytic activity in the serum of a great number of people. They found that healthy men in general have a high protease index, while this was reduced with sick people, particularly patients with chronic inflammations or infections, also in the old age. The serum of cancer patients, of those with precanceroses and patients who later developed malignancies showed a significantly decreased proteolytic enzyme level. The *Gaschler* test is very simple. It is not dependable, but in our experience with over 1000 tests, it gave some valuable information, particularly as a negative exclusion test for malignancies.

Gaschler already started in 1948 taking up again the enzyme therapy of cancer. The events of the war forced interruption of his efforts so that his preparation Carzodelan, which represents, according to his information, a mixture of nucleases with chymotrypsin, was not available again before 1948.

It was clinically applied that time to cancer patients with tumors of all different types, even in advanced stages. With the oral application of his enzyme mixture *Gaschler* could accomplish only some slight local beneficial effects. Therefore he confined his therapy exclusively to parenteral application.

In several clinics of the *Charité* in Berlin, encouraging results were accomplished. Elevation of general well-being, improved appetite, gain in weight and other subjective improvements were registered. With a number of patients tumor regressions could be determined.

All these results were taken in consideration during our own developments and experiments.

We developed in numerous animal experiments and trials at the Biological Research Institute in N.Y. individual enzymes and enzyme combinations with and without activators which had selectively a lytic action on cancer cells.

With the *Maximow*-coverglass method we let grow in cell cultures cancer cells next to normal tissues. Each time 1 mm^2 normal tissue was planted in close juxtaposition to cancer tissue of the same size on coverglasses which were covered with chicken plasma (clot) or collagen substrate. The balanced salts solution (culture medium) added to the control groups was replaced by enzyme solutions with the experimental groups.

Considerable technical difficulties arose first by the fact that the chick fibrin clot as well as other protein-containing substances used to attach the tissue pieces to the glass were liquefied by the proteases added to the cultures, resulting in a floating off of the tissues. Only the weakest enzyme concentrations could be used. Therefore it often required a relatively long time till all cancer cells were destroyed, mostly dissolved. The healthy cells kept on growing normally.

In the control cultures a more or less pronounced interference by the tumor cells could be observed, depending on the types of tumor and normal tissue cells. The cancer cells grew first infiltratively into the front rows of the normal cells and pressed them gradually back. After a few days the normal cells showed necrobiotic signs like vacuolization, pyknosis, some necroses and cytolysis while the cancer cells grew without interference.

In the experimental groups, on the other hand, the following events could be observed: First the cancer cells grew without restraint into all directions, fastest into the direction against the normal tissue. Almost suddenly the cancer cells stopped growing further. They changed partly into a spindle shape, also ball-like form, some shriveled, became enucleated and finally dissolved, while the normal tissues showed hardly any influence by the enzymes added to the cultures. They rather pushed back the front rows of the tumor cells. The cell damages were much more pronounced than those seen in damage to normal tissues observed in the controls.

These characteristic pictures of cell cultures show the reaction of normal and malignant cells under influence by proteolytic enzymes. While fibroblasts (picture 1 and 2) remain uninfluenced, cancer tissue undergoes lysis after a short period of time.

Picture 1 and 2: Normal fibroblast culture 2 and 24 hours after 15 gamma of the enzyme mixture were added to the culture media. Normal growth of these cells. (See pages 223-229)

Picture 3; Malignant tissue (lymphosarcoma) shortly after adding 15 gammas of the enzyme mixture to the culture media: normal growth.

Picture 4: Same culture as picture 3, 2 hours later.

Picture 5: Same culture as picture 3, 6 hours later. Several enucleations took place (round forms).

Picture 6: Same culture as picture 3, 12 hours later. All cells have died due to lysis, intensive disintegration.
Picture 7: Same as picture 6 but more enlarged. Only nuclei are visible, cell membranes and cytoplasma has been dissolved.
Picture 8: Same culture as picture 3, 24 hours later, total lysis.

The enzymes tested in these experimental groups were solutions of trypsin, chymotrypsin, plasmin, kathepsin, pepsin, liver catalase, papain, ficin, bromelin and enzyme extracts of lens esculenta, pisum sativum, aspergillus oryzae, spleen, thymus (mainly nucleases), liver and primarily the enzyme combination finally determined by us as optimal.*

During our investigations it was necessary to develop new tests for assaying the proteolytic and fibrinolytic activity in body fluids. The specificity and sensibility of the plate tests were not high enough (methods after *Astrup* and *Mullertz*). The hemoglobin plate test is a modification of the Anson test and was derived from the old hemolysis assay for differentiation of the different streptococci. Our test is as follows: Hemoglobin is denaturated in alkaline solution by the presence of urea. Through the influence of proteases parts of the hemoglobin molecule are split off causing destruction of the molecule which consists. if intact, of four heme molecules bound to globulin. The splitting by proteolytic enzymes breaks the chromophor complex and the substrate is decolorized.

In casein- or hemoglobin-agar plates, holes of 6mm diameter are punched out and the punched-out circles are sucked up. The serum drop or a solution containing proteolytic enzymes in the punched out holes widen them according to the amount of proteases, which gives an exact indication of the proteolytic and fibrinolytic activity. The assay is not only useful for known proteases, but also for enzyme mixtures and rare proteases from plants and fungi. For a number of various proteolytic enzymes, a straight line is obtained when the log of the product of the

---

* Proteolytic enzymes of fractionated hydrolysates of
   Beef pancreas
   Calf thymus
   Pisum sativum
   Lens esculenta
   Papayotin
   Mannit

two diameters is plotted against the log of the concentration of the enzyme solution. Enzyme solutions, differing in concentrations give parallel dilution curves. The method is also useful for the determination of inhibitor concentrations in biological fluids. (*Maehder, Weigelt*).

In extensive animal experiments on rats we tested the proteolytic activity of the serum after taking the proteolytic enzymes mixture. It was given the animals in gradually increasing amounts orally, intramuscular, intraperitoneal or by rectum. After 90 minutes the proteolytic activity of the serum was determined by the plate method mentioned. A significant relation between the proteolytic potential and the concentration of the enzyme mixture given was shown.

Some authors discuss the importance of general membrane defects in cancer. (*Hoelzl-Wallach*). They distinguish between:

1. plasma membrane (cell contact, cell surface, immunologic changes),
2. mitochondrial membrane (protein and lipid synthesis),
3. lysosomal membrane,
4. nuclear membrane and
5. the endoplasmatic reticulum, responsible for enzymatic changes and enzyme biosynthesis.

It concluded that the membrane hypothesis of tumors postulates that an oncogenic agent acts to introduce an inappropriate protein into or through cell membranes—either in replacement of or in addition to normal components.

Sag'roglu could show in his experiments that in stained malignant tissues many epitheloid cancer cells show smaller or larger membrane defects near the nucleus. Through these tears the cytoplasma leaked out producing the picture of a "nucleus halo" after staining. Such damages of the cell membrane of malignant cells could give a natural explanation for their selective destruction by enzymes.

Exact observations of the cancer cells made it probable that their cell membrane, in contrast to that of fibroblasts and other normal cells, is permeable for the proteolytic and lipolytic enzymes in our mixture. Thus the catabolic enzymes penetrate into the inner cell and are able to dissolve the cytoplasma. The fact that normal cells are more protected against lysis than cancer cells through enzyme inhibitors certainly also plays a part. But since the enzyme mixture used by us contains also lipolytic enzymes against which no inhibitors have been found so far, the cancer cell membrane seems to be insufficiently protected against these enzymes; a factor which therapeutically is very important.

All cell membranes consist mainly of phospholipids and mucopolysaccharides. They cannot normally be attacked by the specific enzymes present in small concentration in the blood, since no sufficiently wide pores exist which would allow the entrance of the macromolecules of the enzyme. The protective cell wall becomes, however, penetrable for lytic enzymes when marked irritations lead to cell damages or to necrobiotic or necrotic processes. In such cases the penetration succeeds easily, also enzymes are set free from the lysosomes, thus bringing about an endogenous lysis.

Electron microscopic and isotopic investigations showed that catabolic enzymes penetrate also through membranes of malignant cells. Possibly this is brought about by the fact that the cell membrane shows defects and that it is incomplete during the rapid mitoses. Perhaps also the cell membrane at the dividing site of the two daughter cells during the telophase of cancer cells is hardly formed for a short time and thereby protects only incompletely the cytoplasma. This would also, in principle, go parallel with the observations of the perinuclear halo of *Sagiroglu*.

In further comprehensive experiments we investigated the effects of the enzyme combination besides in vitro tests, upon the different tumor implants, in rats and mice on chemically induced rat tumors and on spontaneous mammacarcinomas of dogs. Also the preparation Carzodelan® of *Gaschler* was included in the investigations. In both cases definite significant damages of the cancer cells could be demonstrated without influencing normal tissues, although the Carzodelan proved considerably weaker than the enzyme mixture mentioned.

The much lower effectiveness in the animal tests and also later on in the clinical tests on cancer patients treated with Carzodelan® had its cause in the existence of the enzyme inhibitors mentioned above which are found in the serum of cancer patients in individually very differing amounts.

The de-activator is present in all tissues except the thymus.

With cancer patients this de-activator is not found in the tumor and in tumor-bearing organs, but in tumor-free organs in the form of a lactone. In this form it cannot de-activate the applied protective substance (inhibitor); it is biologically inert.

In the solid *Ehrlich*-carcinoma of the mouse, after intratumoral injection of Carzodelan® or of our mixtures, statistically significant tumor regression took place. The tumors partly ulcerated or necrotized, in other cases a partial shrinking in size appeared. Later on the disease process lead to death but the prolongation of life was significant.

By the in vivo systemic application (i.m., i.p. or s.c.) the tumors are influenced only little or not at all, because probably the "lymphocytic membrane" of the solid *Ehrlich* carcinoma is shutting off the incoming vessels as well as the tumor against the enzymes. Also inhibitors which are present in the tissue and in the blood and whose amount and activity is different with every individual, certainly influence the enzyme activity on experimental tumors.

Similar results like with Carzodelan® were seen with our enzyme mixtures. The intra-tumoral use showed marked results, the systemic application less (41, 42). However, the effect on the tumor after systemic enzyme therapy was considerably better if at the same time cortison or cytostatica were given which dissolved that lymphocyte membrane. Then marked regression of tumors and, cessation or slowing down of growth were seen.

All these investigations and test results lead us, as a matter of course, to the conclusion that next to the active enzymes in our mixture, such substances also are desirable which are able to inhibit or block the highly active inhibitors in the cancer serum. The clarification of this item has been the aim of the author (*Wolf*) during his close cooperation with *Freund* (1936-1938) after the intimate connections between inhibitors and anti-inhibitors were recognized. During the examination of a whole series of substances with anti-inhibitory effects, whereby the previously described cell-culture method or also chemically induced or spontaneous animal tumors were used, different substances proved suitable. Among them mainly phospholipoid and lipoproteid combinations which could be extracted out of several glandular tissues. But also mixtures of amino acids, di- and polypeptides, oxidoreductases of the liver, plant polysaccharides and lipoids seemed to be useful for the desired aim.

During the examination of calf pancreas extract it was found that pure trypsin and chymotrypsin are to a great extent inhibited by the cancer serum, but their inhibition is blocked or is absent if amylases and lipase, but also certain other substances of the pancreas extraction are present.

The cytolytic effect of our enzyme mixture was demonstrated by animal experiments of several investigators. The enzymes administered via various routes to hamsters with cheek pouches of hetero-transplantable human tumors showed this antitumor effect (*Goldenberg*, personal communication).

One interesting model for the selective effect of proteolytic enzymes upon tumor cells is the spontaneous mamma adenoma of the Sprague-Dawly rat. In all female animals of this certain strain a spontaneous fibroadenoma resp. adenofibroma develops in advancing age. These tumors develop subcutaneously and can grow into all regions of the body. They may reach a size twice the size of the whole rat. If the proteases are injected intratumorally in these rats, necrosis resp. liquification of the tumor takes place till the entire tumor has disappeared. With tumors up to the size of a hen's egg, these results are always reproduceable, with tumors beyond the size of a man's fist the success is not always constant. As an explanation may be mentioned that the tumor necrotises rapidly in rats and the death of the rat is caused by the overwhelming floodings of the organism with the catabolic products of the tumor. It is remarkable that the enzyme activity stops causing necroses as soon as all tumor tissue is dissolved, the surrounding healthy tissues (connective and muscle tissues) are not affected. Also the necrotized skin over the tumor heals during enzyme therapy without complications. Mostly not even noticeable scar tissue remains (*Weigelt*).

Especially clearly the protective effect of the enzyme mixture can be shown with the sarcoma 180 of the mouse. This sarcoma has a taking rate of over 95%. But in mice which received 4 days before and during transplantation 5 mg of the enzyme mixture, in only 20% of the animals tumors were formed.

REFERENCES

Bainbridge, W. S., N.Y. Med. J. 85, 385, 1907
Beard, J., Enzyme Therapy of Cancer, Schatto etc., London, 1911
Blumenthal, F., Z. Krebsforsch. 10, 137, 1910
Campbell, I. T., J. Amer. med. Ass. p. 1030, 1907
Christiani, A. von, Z. Krebsforsch. 47, 176, 1938; Enzymologica 28, 163, 1965; Enzymologica 28, 235, 1965; Enzymologica 29, 11, 1965; Enzymologica 34, 162, 1968
Cleaves, H. M., Med. Rec. 70, 91, 1906
Curtefield, A., Brit. Med. J., Oct. 1907
Duprey, H., New Orleans Med. a. Surg. J. July 1907
Freund, E., Wien. med. Wschr. 12, 1934
Freund, E., Wien. klin. Wschr. 46, 1576, 1933
Gaschler, A., Parenterale Fermenttherapie maligner Tumoren, Ulm, Haug Verlag, 1961
Goeth, R. A., J. Amer. Med. Ass. 1907
Hald, P. T., Lancet II, 1371, 1907
Kretz, J., Wien. klin. Wschr. 6, 1936
Little, W. L. A., J. Amer. Med. Ass. page 1724, 1908
Marsden, A., Gen. Prac. 22, 1908
Physick, A., etc., Pharmacotherapeutics, D. A. Appleton, N.Y. 1928
Pusey, W. A., J. Amer. Med. Ass. 46, 1763, 1906
Ransberger, K., Sigma Xi Conference, Colorado State University, 1967
Shaw-McKenzie, J., Brit. Med. J. 1, 715, 1906
Sagiroglu, N., Amer. J. Obstetr. a. Gyn. 85, 454, 1963

CANCER THERAPY WITH SPECIFIC ACTING ENZYMES

Two basic possibilities exist for the treatment of cancer with specific enzymes:
1. Certain quantitative deviations in the tumor metabolism are utilized, as in the case of arginase.
2. Individual special deleting phenomena are used interfering with the metabolism of tumor cells (for instance, asparaginase).

Arginase

Since arginin furthers the growth of tumors, in theory inhibition of the tumor through enzymatic destruction of the essential amino acid arginin is to be expected. If one administers highly purified arginase to a cell culture of fibroblasts with *Jensen*-sarcoma cells, the inhibition of the tumor growth runs parallel with the activity of the enzyme. Enzyme preparations inactivated by heat have no influence upon the mitosis rate.

One can observe complete remissions of the *Walker*-carcinoma of rats within 3 to 10 days after the injection of arginase.

If the arginase is administered intraperitoneally, the plasma enzyme level rises within one to two hours. This, incidentally, also proves the absorption of the enzyme from the peritoneum into the blood stream.

Unfortunately the effect of arginase is very quickly exhausted since the host organism forms anti-bodies against the foreign protein.

The treatment of malignant tumors with arginase cannot as yet be tested on many cases since no industrially manufactured preparations are available. Furthermore, the degree of purity and the production method doubtless has an influence upon the effectiveness.

Asparaginase

The L-Asparaginase (L-Asparagin-aminohydrolase) breaks down enzymatically the amino acid asparagin to asparaginic acid and ammonia.

Some tumors and leukemia cells cannot synthetize asparagin by themselves, and are therefore dependent upon an exogenous supply, for instance through food. Asparagin thus presents for these tumor cells an essential amino acid without which they are not able to mitose and proliferate. If L-Asparaginase is supplied to tumor cells, it destroys the amino acid asparagin. The intracellular metabolism of the tumor cells is thereby inhibited at some still unknown point and its cells perish (*Broome* and others).

Most normal cells synthetize sufficient asparagin and do not depend upon an exogenous supply, so that asparaginase does not influence their growth.

The L-Asparaginase is the first enzyme-based cytostaticum, it inhibits, in vitro and in vivo, the growth of some tumors.

The story of Asparaginase research begins in 1953, through a chance observation. *Kidd* noted in his tests that guinea-pig serum given i.p. slows down the growth of some tumors in mice and rats. *Broome* proved later that the active factors against cancer in the serum of guinea-pigs is identical with L-Asparaginase.

The serum of newly born guinea-pigs contains only a little L-Asparaginase and therefore shows only a minute anti-tumoral effect. Among the entire animal world only the serum of the guinea-pigs and of a South American steppe hare (*Agouti*) contains larger quantities of asparaginase.

The site of formation of the serum asparaginase is not known; and an explanation for the presence of the enzyme just in the serum of *Agouti* and guinea-pigs can not be given for the time being.

L-Asparaginase can be extracted in purest form from the serum of guinea pigs. More rewarding, however, is the biochemical production from *Escherichia-coli* cultures which are also rich in L-Asparaginase.

The enzyme activity can be determined by measuring the released ammonia.* L-Asparaginase from coli cultures has a half-time value of approximately 24 hours in the human serum.

The enzyme is not eliminated through the kidney and likewise does not pass through the blood-liquor barrier. According to current research, carcinoma cells lack the ability of L-Asparagin synthesis.

Clinical research in the use of L-Asparaginase can, up to now, be found only in a few publications. Chief indications are the hemoblastoses. Unfortunately, the first successful results of the treatment were subsequently hindered to a large degree by the appearance of resistance against the enzyme.

In order to reach a final judgement regarding the therapeutic value about this new enzyme therapy, further observations are required and critically evaluated. Here it may only be mentioned that the use of Asparaginase is connected with certain criteria, like freedom from pyrogens of the preparation, sensitivity of the tumor tissue etc.

---

* International unit is the enzyme quantity which releases 1 micromol of ammonia from asparagin in 1 minute.

PATHOLOGY AND BIOCHEMISTRY OF METASTASIS FORMATION

The formation of metastasis of malignant tumors in the organism was subjected to intensive research only in recent years.

Cancer researchers agree that the process of metastasis can be sub-divided into several phases:

1.) The shedding of individual or groups of cancer cells from the primary tumor and their entering into the smallest vessels.

2.) The transport of these cells through the blood or lymphatic system.

3.) The settling of cancer cells in organs and

4.) their proliferation under suitable conditions with succeeding autonomous growth of a metastasis.

(*Willis* (1), *Zeidmann* (2), *Schmähl* (3, 4), and others).

It took almost a century till a conception of the mechanism of metastasis based on experiments was formed. *Thiersch* (5) described in 1865 the spreading of carcinoma cells through the veins and *Ashworth* (6) traced anomalous cells and cancer cells in the blood of patients who had died from cancer. *M. B. Schmidt* (7) was the first to indicate in 1903 that cancer cells cling to the endothelium of pulmonary capillaries and are surrounded by a fine net of fibrin, in which leucocytes and thrombocytes are lodged. Even the smallest tumors may transfer cancer cells into the blood stream, as proved by *Gastpar* (36). He discovered among a total of 162 cases of carcinomas of the vocal cords of rice kernel size, cancer cells 90 times in the peripheral venous blood already at the first examination. The shedding of the tumor cells is a continuous process and there is no appreciable difference whether it happens via blood or the lymphatic path.

Lymph nodes are no handicap since they can be traversed by the individual cells or cell units in the shortest possible time, *Fischer* (11)). Also the capillary areas of the liver, kidney, spleen, mesentery and lungs present no barrier to the extension of cancer cells. Capillaries and walls of larger vessels can also be penetrated by even very large cancer cells (*Zeidmann* (2)).

A malignant infiltration may, however, also occur without

shedding of cancer cells from the primary tumor. By the tension which prevails in the growing tumor cancer cells lying on its periphery, they are forced into the surrounding tissues and will thus infiltrate them (*Willis* (1)). Such cells show fibrin formation on their cell membranes; they are able to damage or dissolve peripheral soma cells and to take their place. It is probable that the invasive growing cancer cells absorb the lysed products of the destroyed soma cells for their own metabolism and that thereby a stimulation of their own mitosis occurs. This process explains the stationary infiltrating growth of malignant tumors without spreading of the cancer cells by the blood or lymph paths.

The flushing off and extension of the cancer cells over the communication system of the organism has been proved by many later examinations and our knowledge gained was thus enhanced. (*Baserga* (12), *Iwasaki* (13), *Saphir* (14), *Takahashi* (15), *Walther* (16), *Willis* (1) and *Wood Jr.* (17)).

Cancer cells may also circulate in the blood for a lengthy period without adhering to any spot. This was already pointed out by *Ambrus* (18) 20 years ago by the results of his experiments with radioactively marked ascites tumor cells.

Roberts (10) also assumes that cancer cells may circulate in the blood and are thereby either destroyed spontaneously or they form metastases after settling somewhere. From his experiments he concluded that circulating tumor cells which do not settle down will not create a metastasis. It has been proved by experiments with animals and in cell cultures that such circulating cells are viable.

Agostino recapitulates the problems of metastasis formation (21) and his research also indicates that cancer cells must be present in the blood or lymph system in order to be able to produce a telemetastasis. While a large part of these cells is destroyed, whereby certainly the lytic breakdown through serum proteases plays an essential part, some cancer cells which are still intact are anchored in the vascular system at the endothelium of the vessels (e.g. the lungs and other organs) and there induce a metastasis. Thus relation between the vascular endothelium and the cancer cells represents the central problem of metastasis development, since the adhering and "sessile" cells on the endothelium now grow very fast and infiltrate the vascular wall (*Levin* (23), *Jonasson* (24)). Tumor cells emigrating from the vessels may however also wander through the intercellular tissues and on their way through new lymphatic vessels return again into the blood circulation. This recirculation was also shown for lymphocytes and other blood elements (*Gowans* (69). Also the cells adhering to the endothelium of the vessels can be destroyed spontaneously (*Roberts* (61), (*Levin* 23)) or may remain inactive, perhaps for many years (*Hadfieldt* (25), *Fischer* (53)). In American literature this is called "dormant state".

The causes for a re-activation of these cells are not completely known, they may depend upon many endogenous and exogenous factors. Changes in the hormone balance (*Williams* (26)), stress after an operation, injuries, cancerogenous compounds and physical influences (for instance sun rays) are mentioned as possible causes for a re-activation of the dormant cancer cells.

Aside from the reactivation of the cancer cells, such factors also contribute to the faster growth of the primary tumor. (*Long* (27), *Schatten* (28)).

The invasion of malignant tumors into the blood stream and the fixation of the cells to the vascular endothelium are influenced by physicochemical factors. The tumor cell, as an independent living unit, has its own surface characteristics which only lately were subject to closer examination. The results of this research have materially contributed to the understanding of the metastasis problems. Cancer cells have an enhanced mobility (*Abercrombie* (29) and a reduced cohesion or affinity to each other (*Coman* (37)). The hyaluronidase secreted by carcinoma cells ("spreading factor") is supposed to promote the shedding and settling down of the cells. In this connection newer physicochemical investigations of the electrostatic potential of the cancer cell in comparison with normal homologous cells are of interest. *Purdom* has demonstrated that an increased malignancy of the cancer cell is accompanied by an altered higher electro-negative charge of these cells (32). Cataphoretic research of the cells indicated also that the electric charge of various human tumor cells differs from that of the normal mother cells. (*Ruhenstroth-Bauer* (33, 34)).

Parallel to it are the results of investigations which show that the adhesion of blood cells to the endothelial cells is prevented by the same electro-negative charge of both types of cells. Changes in the electric charge of the membranes can be produced experimentally. After the feeding of the carcinogen dimethylnitrosamine to rats, all their liver cells were found to have a higher negative charge than liver cells of untreated animals. The morphological mutation of the liver cell into a carcinoma cell is a secondary phenomenon which is perhaps conditioned by the change of electric potential of the liver parenchyma (*Ruhenstroth-Bauer* (34)).

If one reviews the facts just mentioned he must deduce that metastasis formation is closely connected with the relation of the cancer cell to the vascular endothelium and that morphological and physicochemical mutual relations must exist between the two.

The cancer cell, surrounded by a fibrin net, in which, in addition, leucocytes and thrombocytes are embedded, reacts as a foreign body to the endothelium, with which it must come to terms in one way or another. Histological investigations show that the endothelium loses its normal structure if a cancer cell is attached as a wall-thrombus.

Wood (35) recapitulates the metastasis process by considering his own research and that of others as follows:

"*Cancer cells can only form metastases if they adhere to the vascular endothelium, for in the animal experiment as well as in the human being, the endothelium is the always present dynamic host for the circulating cancer cells and also the barrier raised against them.*"

As in other medical problems, the exact proof of living cancer cells in the blood stream was depending on the development of modern bio-techniques. W. S. *Schmidt* and other researchers used fixed and stained histological sections for this purpose. The mutual effects between vessel endothelium and cancer cell however are of a dynamic nature and operate between vital cells and cell compounds whose behavior is influenced by the speed of the blood stream etc. Therefore it is not entirely improbable that the thrombus observed with tumor cells in the section was developed after the fixation and does not represent the primary conditions.

Only intravital microscopy and cinematography, through which the behavior of the tumor cells can be shown in vivo in the blood stream and observed in the early stages, furnished the final proof for the behavior of cancer cells in the blood stream. By this method a number of researchers were in a position to recognize clearly that individual, living tumor cells or cell compounds adhere to the endothelium, are sometimes torn away again and circulate anew in the blood. At the fixation of the tumor cells on the endothelium a parietal thrombus is formed within 20 minutes which includes the tumor cells, leukocytes and lymphocytes surrounded by a fine fibrin net (*Wood* (35), *Gastpar* (36), *Zeidman* (2)).

The normal epithelium loses its original structure and is infiltrated by lymphocytes and leucocytes. Small vessels can thereby be completely obliterated in a short time by thrombi. The following diagram shows these observations:

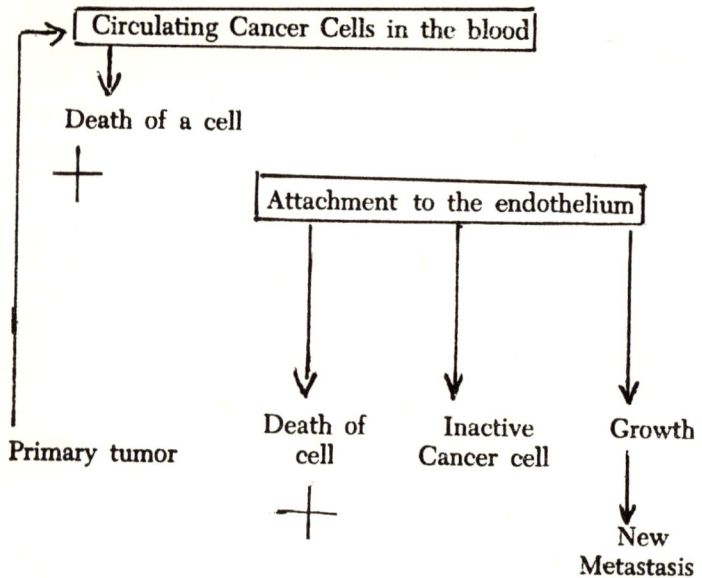

The adhesion and what United States scientists have coined "cancer cell stickiness" are different physical processes, respectively characteristics of the cell. They have their cause in tumor-specific physicochemical surface conditions of the cells (*Abercrombie* (29), *Coman* (30), and others).

By cohesion we understand the attraction of equal cells to each other, while in the "stickiness" the affinity or adhesion of the cell to different substrates (for instance foreign cells, glass plates) is increased. Cancer cells are therefore extremely sticky and ready to adhere, while normal cells are characterized by a strong tendency to cohere and by only slight adhesion. The cells of various animal tumors have a specific, greatly varying stickiness to glass. The degree of stickiness is decisive for the rate of metastasis. The tumor cells of the ascites hepatoma AH 13 are more sticky (+ 54%) than the cells of the ascites hepatoma AH 7974. On account of the increased stickiness, the rate of transplantation with the AH 13 tumors is five times higher than that of the AH 7974 tumor (*Kojima* (63)).

Permeability, stickiness and adhesion are surface factors whose importance for the start of a tumor was discussed at the 10th International Cancer Congress in Houston, Texas 1970.

Todorutiu et al. demonstrated in their experiments the importance of a preliminary treatment of cancer cells before their implantation. Wistar rats received 1 mill. Walker tumor cells/ml which were incubated before this transfer either in physiological salt solution or in a 1% Tween-80, solution. While in the control group only 10% of the animals showed tumors, in the Tween group it rose to 46.6%.

The influence upon the stickiness of ascites tumor cells by some chemical substances with which the tumor cells were pretreated, shows the following table, according to *Kojima:*

The numbers of cells remaining on the glass plate are expressed in percentage to the controls (= 100%).

	0	10	30	50	70	100%
Protamine sulphate		49.6%				
Tween 80					107.8%	
Desoxycholic acid					104.7%	
Trypsin	13.9%					
Calcium chloride					100.2%	

It is remarkable that the proteolytic enzyme trypsin reduces the stickiness markedly. From this we may conclude that this is caused by a surface covering of fibrin. Trypsin dissolves this fibrin proteolytically whereby it is eliminated as the main stickiness-inducing factor.

The fibrin plays a central role in the blood coagulation and the fibrinolysis. Through the cancer cell thrombus, the metastasis is intimately related to these reactions, therefore, the fibrin is once more in a central position of the metastasis process.

Several authors have demonstrated that a number of experimental tumors have a high fibrin content (*Bale* (39), *Agostino* (21), *Day* (40, 41) and *Spar* (44, 45)). *Day* was able to show with rats that tagged fibrinogen is stored in the tumor cell (*Murphy*-Lymphosarcoma). Human tumors also have an increased fibrin content (*Hiramoto* (42)). The *Murphy*-Lymphosarcoma of the rat stores 15 to 25% J-131 fibrinogen after i.v. injection.

Compared with body tissues, the fibrinogen concentration in the cancer tissue is 4 to 15 times higher. However if one treats the test animals with proteases or with heparin or *Warfarin* (anticoagulant), the fibrinogen storage in the tumors is reduced by 60 to 80%.

Autoradiographic tests on rabbits, which were given radioactive fibrinogen and a supension of V_2 cancer cells showed a fibrinogen accumulation in the microthrombus which had formed around the cancer cells adhering to the endothelium. But only the endothelium-adhering tumor cells absorb fibrinogen (*Wood* (35)), while the intravascularly circulating cancer cells have no affinity to the fibrinogen and die without forming metastases (*Wood* (35)). The following diagram shows the storing of fibrinogen and cancer cells:

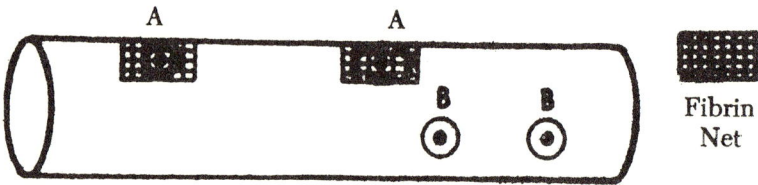

A = adhering cancer cells with thrombus and fibrinogen storing.
B = free cancer cells in vascular lumen, no storage ability, death of cells.

From all investigations which have been reported up to now, it appears that the cancer cell attaches itself to the vessel wall and induces a microthrombus there. It follows that the thrombosis process should be influenced by the action of fibrinolytica or anticoagulantia, since they induce thrombolysis or fibrinolysis. In 1953 *Lawrence* (47, 48) showed for the first time that a preliminary treatment with heparin reduces the success rate of transplants of experimental tumors by 70%. All rabbits, which had received intravenous injections of V_2 carcinoma cells died due to pulmonary thrombosis, if not protected with heparin.

5 mg/kg bw. heparin, given before the inoculation of the cancer cells, prevented any lung emboli. If the heparinisation was continued for 60 days, fewer than 20% of the animals developed lung tumors, while 100% of the surviving controls showed tumors (51).

Analogous tests with plasmin and cancer cells are described by *Cliffton* (52, 55, 56). If experimental animals are treated with plasmin before the injection of cancer cells, a complete protection against a fatal lung embolus takes place, at the same time a reduction of the transplantation rate is connected with it. The cancer cells circulate in the blood, however the attachment to the endothelium and the formation of a wall thrombus is prevented. Plasmin acts fibrinolytically, it attacks the micro-thrombi and thus deprives the carcinoma cell of a basis for its growth and metastatic tumor formation. These facts have been confirmed by many scientists, for instance by *Wood* who clearly put forward the protective result of fibrinolytica and anticoagulantia in this connection (*Agostino* (46), *Fisher* (53), *Koike* (54), *Wood* (35), *Grossi* (49, 50) and others).

The following summary shows the most important experimental publications on this work:

Tumor	Test animal	Protective substance	Trans-plantation	Author
V 2 Carcinomsarcoma	Rabbit	Plasmin	i.v.	Clifton
V 2 Carcinomsarcoma	Rabbit	Plasmin	i.v.	Wood
Brown-Pearce Carc.	Rabbit	Plasmin	i.v.	Clifton
Walker Ca. 256	Rat	Plasmin	i.v.	Clifton
Walker Ca. 256	Rat	Plasmin	i.v.	Grossi
Walker Ca. 256	Rat	Heparin	i.v.	Agostino
Walker Ca. 256	Rat	Heparin	i.v.	Fisher
V 2 Carcinoma	Rabbit	Heparin	i.v.	Lawrence
Lewis Sarc. 241	Mouse	Heparin	i.v., s.c.	Wood
Lewis Ca. 150	Mouse	Heparin	i.v.	Wood
Yoshida-Sarcoma	Rat	Plasmin	i.v.	Hiemeyer
Yoshida-Sarcoma	Rat	Wobe-Mugos	i.v.	Wolf

Hiemeyer and *Merkle* report on the checking influence of plasmin upon the starting of implanted tumors. While the results of their investigations point to a fibrinolytic activity as a cause of antitumor activity, a direct cytolytic effect of the plasmin appears rather more probable.

In another publication (62a) the mechanism of activity of plasmin was clarified by viability tests in vitro and in vivo. The vitality of the tumor cells was determined by the method described by *Schreck* (1949) and *Kaltenbach*, according to which tumor cells can be dyed after extinction of their own cell metabolism by means of acid dyes, e.g. trypanblue.

If tumor cells are incubated with inactivated plasmin, no special differences regarding the part of the damaged cells compared with the control cells could be seen which were kept in isotonic salt solution. If freshly prepared—that means fully active —plasmin solutions are used, 24 hours later 100% of the tumor cells could be dyed, as a sign of a cell damage induced by plasmin. If these "dead" cells are injected into rats, all experimental animals survived. Similar results were found with cells which were incubated for a shorter period in high concentrations of plasmin; also here no tumor started. Thus the author confirmed the opinion already expressed by others that plasmin has a cytotoxic affect on tumor cells and that, with it, plays a significant part as a body proper substance for the defense against tumors.

In agreement with these findings are investigations of other authors about the cytotoxic activity of anti-coagulantia. *Dicoumarol* has a toxic effect in vitro upon human cancer cells which runs in proportion to the concentration (*Lionell*). *Bhuyan* proved the direct cytotoxic activity of *Warfarin* in vitro upon leukemia cells (Mouse L 12 10).

Thornes investigated the influence of anticoagulantia upon the mobility of V_2 carcinoma cells in vivo by means of the chamber method on the rabbit ear. The mobility of the cancer cells and other body-proper cells is followed up kinematographically over a certain time unit. (Details about the method found e.g. in *Wood, Jr.*, J. Soc. Motion Picture and Television Engineers 74, 737, (1965) and S. *Wood*, etc. Bull. Hopkins Hosp. 119, 1 (1966)).

Rabbits were treated previously for 4-5 days with *Warfarin* and received thereafter 10^6 V_2 carcinoma cells. The motility of the cells followed up over 24 days with a mean photo-time of 240 hours. With the animals pretreated with *Warfarin* the motility was reduced still about 20 days after pretreatment with anticoagulants. Granulocytes, lymphocytes and macrophages were not influenced.

Dicoumarol and *Warfarin*—probably also other anticoagulantia—act directly upon the cellular metabolism by checking the oxydative phosphorylation (*Martius*). The inhibition of the cell motility by anticoagulantia could be connected with this fundamental bio-energetic reaction. The absorption of Dicoumarol by isolated liver-cell mitochondria is known, corresponding investigations about the reaction of carcinoma cells are still expected (*Howland*).

Wolf and his collaborators also worked with experimental tumors. They used as fibrinolytically active substance the combination of animal and plant proteases (Wobe-Mugos). The results on the test animals were similar.

Approximately 100,000 cells of the *Yoshida*-carcinoma were implanted into 8 months old rats. Two hours before or together with the transfer of the tumor cells, the animals received 20 mg Wobe (i.p.). This enzyme treatment was continued for 12 hours every 2 hours. Of 120 animals represented in the test, 70% survived, while of the untreated controls only 5% remained alive. The average time of survival of the treated animals was 29 days, that of untreated animals was 22 days.

A comprehensive discussion about the biochemistry of metastasis formation, importance of the enzyme level for the start, therapy and prophylaxis of tumors and the connection between tumor growth and Vitamin A was given by *Ransberger* at the 10th International Cancer Congress, Houston 1970.

The rate of metastasis of tumor cells in the blood stream can be increased through stress (*Wood* (35), hyperlipaemia (*Wood* (35), *Cliffton* (52), cortisone derivatives, (*Wood* (35), growth hormones, *Wood* (35) and endotoxins, (*Wood* (35)). According to *Eichenberger*, endotoxins shorten the blood coagulation. Also exogenous damages (trauma, hypoxemia) of certain organs and vessels bring about an increased deposit of cancer cells upon the endothelium and thereby favor metastasis formation.

It is conceivable that due to trauma the endothelium of the involved capillary area undergoes such a change that the local coagulation rate of the blood becomes impaired. However, by supplying anticoagulants or fibrinolytica before or soon after the trauma, the increased tumor cell affinity to the damaged endothelium is again reduced.

The treatment of tumor cells with nitrogen mustard increases their ability of adhesion to the endothelium and the rate of transplantation, (*Cliffton* (56)). *Cliffton* proved that the transplantation rate of 50,000 V_2-Carcinoma cells, given i.v., is 100%. If one uses pre-irradiated tumor cells, it drops to 80% and after pre-treatment of the animals with plasmin for several days sinks to 10%.

Strong X-ray doses, however still under-threshold, damage both the endothelium as well as the cancer cell. Their stickiness is increased and thereby also the adhesion to the endothelium (*Wood* (35)). If the test animal is given at the same time healthy and damaged cells i.v. the frequency of metastasis is increased (*Wood* (35), *Seelig* (64)).

Perhaps these observations will furnish an explanation for the fact that by irradiating a primary tumor sometimes a metastasis is produced (*von Essen* (65)).

The result of anticoagulants upon the rate of metastases was only recently again confirmed. *Hiemeyer* was able to show that a continuous plasmin infusion (over 22 to 24 hours) given at the same time with the i.v. injection of *Yoshida*-carcinoma cells could prevent the formation of this tumor in 52% of the animals while only 8% of the controls remained tumor free.

Since anticoagulantia and fibrinolytica reduce the metastasis quota, it should be possible also to increase it by specific inhibitors. For such experiments inhibitors are available today (59.58) through extractions of tissues (A) and synthetic substances (B).

Rats, which receive 25,000 *Walker* carcinoma cells i.v. and fibrinolysis inhibitors, show a significant rise of lung metastases compared to the controls, (*Cliffton* (52), *Boeryd* (58)).

The table below shows the rise of the lung metastases per 100 animals after receiving natural (A) or synthetic (B) fibrinolysis inhibitors.

Test Groups				
Controls	20 mg B	40 mg B	40 U A	75 U A
30%	39%	85%	43%	72%

From these results we may assume that the cancer cells, under the influence of the fibrinolysis inhibitors, possibly due to increased stickiness, can be glued faster and more lasting to the endothelium, and the fibrinolytic effect of the tissues and blood locally at the site of the thrombosis prevented. Similarly the surrounding conditions in the vascular system could be changed by the inhibitors in favor of the carcinoma cell.

These tests furnish a very impressive example for the importance of biochemical reactions in pathological processes. They indicate, moreover, that in the complicated co-existence of enzymatic processes, physical factors etc., the fundamental reactions: fibrinolysis—blood coagulation—hemodynamic equilibrium are just as valid for the metastatic processes.

Observations and tests dating back 100 years can now be explained and conclusions can be drawn.

Trousseau pointed out already in 1865 that connections must exist between the appearance of a carcinoma and the tendency to thromboses in cancer patients.

From various statistics can be seen the frequent appearance of thrombosis in cancer cases: *Sproul* discovered among 3,258 cancer autopsies 14.4% thromboses of the arteries and veins. Among a great number of autopsies *Thies* (67) found approximately 42% thromboemboli in cancer patients, while only 32% were found in non-cancer cases.

Interesting are the investigations of *Ludwig* who reports on the results of a prophylactic therapy with anti-coagulantia during radiation therapy of genital carcinoma in women in the last 6 years (68).

The prophylaxis with anti-coagulantia was either confined to the times of the application of the radioactive preparation or continued during the interval. As anti-coagulant *Ludwig* chose a heparinoid which was administered as a dosage of 800 to 1200 u/kg body weight within 24 hours. This treatment had primarily only the aim of preventing thromboembolic complications during the radiation therapy (68a, 68b).

Genital-Ca. Stage II and III	Total Number	Prim. Mortality	Prim. recurrences Anticoagul. group	Controls
	2278	1.01%	9.29%	17.99%

The difference in the recurrence index of squamous epithelial carcinoma after radiation therapy has been significantly changed in favor of the group which received during the ray treatment at the same time anticoagulants.

In order to clarify this remarkable and unexpected influence of the prophylactic anti-coagulantia therapy upon the metastasis formation, *Ludwig* followed up by means of a specific dyeing method the fibrination in the terminal circulation of the portio carcinoma. The histological investigations showed that capillaries and postcapillaries were blocked by tumor cells and fibrin deposits and the tumor bulgings were surrounded by broad fibrin seams. The fibrination begins regularly at the side of the capillary next to the tumors. If before the removal of tissue sufficient amounts of an anti-coagulant are given, the terminal blood perfusion of the tumor is considerably improved and only traces of fibrin can be found. Hand in hand with an improved blood transfusion of the tumor tissue is an increased rate of mitosis, which again leads to an enhancement of the sensibility toward gamma rays. By the prevention of the appositional fibrination around the destructively growing tumors by means of heparin the radiation sensibility of the tumor is improved. Tumor cells tear along with them during the hematogenic spread some fine pericellular fibrin coating which favors the stickiness of the tumor cells to the endothelium (70). In the group of heparin-treated and irradiated carcinoma patients this pericellular fibrin within the capillaries is very much rarer found.

At an international meeting on cancer and blood coagulation several authors insisted that in a case of thrombophlebitis of unexplained etiology it should be mandatory to search for a beginning tumor (*Gross, Lasch, Deutsch, Marx, Witte, Gastpar*).

Literature

1. *Willis*, R. A., Spread of Tumors in the Human Body. C. V. Mosby (1952)
2. *Zeidmann*, I., Metastasis. Cancer Res. 17, 157 (1957)
3. *Schmähl*, D., Entstehung, Wachstum and Chemotherapie maligner Tumoren, Editio Cantor (1963)
4. *Schmähl*, D., Experimentelle Grundlagen der Tumor meta-

stasierung, in; Krebsforschung and Krebsbekämpfung, 6, 176 (1967)
5. *Thiersch*, K., Der Epithelialkrebs, namentlich der Haut, Leipzig (1865)
6. *Ashworth*, T. T., Austral. Med. J., 14, 146 (1869)
7. *Schmidt*, M. B., Verbreitungswege der Carcinome, Fischer, Jena (1903)
8. *Cole*, W. H., Dissemination of Cancer, Prevention and Therapy, N.Y. (1961)
9. *Zeidman*, J., Cancer Res. 21, 38 (1961)
10. *Roberts*, S., Ann. Surg. 154, 362 (1961)
11. *Fischer*, B., etc., Cell Science 130, 918 (1959)
12. *Baserga*, R., etc., Arch. Pathol. 59, 26 (1955)
13. *Iwasaki*, T., J. Pathol. Bact. 20, 85 (1915)
14. *Saphir*, O., Amer. J. Pathol. 23, 245 (1947)
15. *Takahashi*, M., J. Pathol. Bact. 20, 1 (1915)
16. *Walther*, H. E., Krebsmetastasen, Schwabe, Basel (1948)
17. *Wood*, S., Jr., Proc. Amer. Ass. Cancer Res., 2, 260 (1957)
18. *Ambrus*, J. L., Ann. N.Y. Acad. Sci. 63, 938 (1956)
19. *Engell*, H. C., Ann. Surg. 159, 456 (1956)
20. *Wood*, S., Jr., Canad. Cancer Conf. 4, 167 (1961)
21. *Agostino*, D., etc., J. Nat. Cancer Inst. 27, 17 (1961)
22. *Agostino*, D., etc., Ann. Surg. 161, 97 (1965)
23. *Levin*, I., etc., Proc. 8, 114 (1910)
24. *Jonasson*, O., Surg. Forum 9, 577 (1958)
25. *Hadfieldt*, G., Brit. Med. J. 2, 607 (1954)
26. *Williams*, R. H., Endocrinology, Saunders, Philadelphia (USA) (1925)
27. *Long*, L., etc., Acta Union contra Cancer 19, 464 (1959)
28. *Schatten*, W. E., etc., Cancer 11, 460 (1958)
29. *Abercombie*, A., etc., Cancer Res. 22, 525 (1962)
30. *Coman*, D. R., Cancer Res. 13, 397 (1953)
31. *Simpson*, W. L., Ann. N.Y. Acad. Sci. 52, 1125 (1950)
32. *Purdom*, L., etc., Nature 181, 1586 (1958)
33. *Ruhenstroth-Bauer*, G., Naturwiss. 49, 363 (1962)
34. *Ruhenstroth-Bauer*, G., etc., Klin. Wschr. 44, 30 (1966)
35. *Wood*, S., Jr., Bull. Swiss. Acad. Med. Science, 20, 92 (1964)
36. *Gastpar*, H., Thromb. Diath. haemorrh., Suppl. 28, 119 (1967)
37. *Coman*, D. R., Cancer Res. 13, 397 (1953)
38. *Agostino*, D., etc., Ann. Surg. 157, 400 (1963)

39. *Bale*, W. F., etc., Cancer Res. 20, 1488 (1960)
40. *Day*, E. D., etc., J. Nat. Cancer Inst. 22, 413 (1959)
41. *Day*, E. D., etc., J. Nat. Cancer Inst. 23, 799 (1959)
42. *Hiramoto*, R., etc., Cancer Res. 20, 592 (1960)
43. *O'Meara*, R. A. M., Arch. de Vecchi abat. path. 31, 365 (1960)
44. *Spar*, I. L., etc., Cancer Res. 20, 1511 (1960)
45. *Spar*, I. L., etc., Proc. Amer. Ass. Cancer Res. 4, 65 (1963)
46. *Agostino*, D., etc., J. Nat. Cancer Inst. 27, 17 (1961)
47. *Lawrence*, E. A., etc., Proc. Amer. Ass. Cancer Res., 1, 32 (1953)
48. *Lawrence*, E. A., etc., 38th Clin. Congress Amer. Coll. Surg., N.Y. (1952)
49. *Grossi*, C. E., etc., Cancer 14, 957 (1961)
50. *Grossi*, C. E., etc., Arch. Surg. 84, 87 (1962)
51. *Lawrence*, E. A., etc., Surg. Forum 3, 694 (1953)
52. *Cliffton*, E. E., etc., Cancer 9, 1147 (1956)
53. *Fischer*, B., etc., Surgery 50, 240 (1961)
54. *Koike*, A., Cancer 17, 450 (1964)
55. *Cliffton*, E. E., etc., Cancer Res. 21, 1062 (1961)
56. *Cliffton*, E. E., etc., J. Nat. Cancer Inst. 33, 753 (1964)
57. *Kraut*, H., Leistungen und Ergebnisse der neuzeitlichen Chirurgie, Stuttgart, p. 62 (1958)
58. *Boeryd*, B., etc., pathol. europ. 2, 181 (1967)
59. *Mellgren*, J., etc., Acta pathol. microb. scand. 68, 535 (1966)
60. *Boeryd*, B., Acta pathol. microb. scand. 65, 395 (1965); 68, 547 (1966)
61. *Roberts*, S., etc., Surg. Forum 8, 146 (1958)
62. *Hiemeyer*, V., Z. Krebsforsch. 72, 36 (1969)
62a. *Schreck*, R., Arch. Path. 37, 319 (1944)
62b. *Kaltenbach*, J. P., etc., Ann. N.Y. Acad. Sci. 63, 977 (1956)
63. *Kojima*, K., etc., Cancer Res., 24, 1887 (1964)
64. *Seelig*, K. J., Brit. J. Cancer 14, 126 (1960)
65. *von Essen*, C. F., J. Natl. Cancer Inst. 12, 883 (1952)
66. *Michaelis*, L., Lancet, S. 832 (1964)
67. *Thies*, H. A., Zbl. Phlebol. 3, 146 (1967)
68. *Ludwig*, H., in: E. Deutsch etc. (editors) Fibrinstabilisierender Faktor, Struktur des Blutgerinnsels, Krebs und Blutgerinnung, F. K. Schattauer-Verlag, Stuttgart (1968)
68a. *Ludwig*, H., Geburtsh. u. Frauenheilk. 22, 1121 (1962)

68b. *Ludwig*, H., München. med. Wschr. 108, 1035 (1966)
69. *Gowans*, J. L., J. Physiol. London 146, 54 (1959)
70. *O'Meara*, R. A. Q., Irish J. Med. Sci., 391, 327 (1958)
71. *Lisnell*, A., etc., Acta Path. 57, 145 (1963)
71a. *Bhuyan*, B. K., cit. with 72
72. *Thornes*, R. D., etc., J. Hopkins Med. J. 123, 305 (1968)
73. *Martius*, G., etc., Biochem. Biophys. Acta. 13, 289 (1954)
74. *Howland*, J. L., Biochem. J. 106, 317 (1968)

CLINICAL RESULTS OF THE ENZYME THERAPY OF MALIGNANT TUMORS

Unexpected Difficulties in Clinical Trials

It appears essential to the nature and purpose of this monograph to allow a broad space for clinical observations and results of the enzyme therapy of malignant tumors. Thereby it cannot be avoided that in the casuistic enumeration always again the same therapeutic results are mentioned which by this treatment were observed by the reporting physician in remarkable uniformity. This may help to prove the efficiency of the enzyme therapy even if statistical exactness is lacking which would be very difficult to perform.

I (*Wolf*) kept myself occupied, since the scientific collaboration with *Ernst Freund* in Vienna started 40 years ago, with the possibilities of enzymatic influence of malignant tumors. This was suggested to me partly by the interesting work of the Scottish embryologist *John Beard*, partly by the discoveries of *Ernst Freund* and his assistant *Kaminer*, who showed that in the serum and urine of cancer patients an enzyme is missing which exists in the serum of healthy people and which is able to dissolve malignant cells in vitro. My collaborators and myself started after many years of laboratory tests and animal experiments 25 years ago, to use proteolytic enzymes regularly in the treatments of tumor patients. Right from the start I realized that such therapeutic work would not suffice for statistical evaluation because it was for us entirely impossible to perform control-blind or even double-blind investigations, much less one at random therapy.

The patients were referred or came to me by their own decision to be treated with proteolytic enzymes. It would be impossible and could neither ethically nor legally be justified to treat such patients perhaps with placebo preparations, only in order to satisfy scientific statistics.

Besides the greater part of the cancer patients came to us in such a far advanced stage that curative therapeutic endeavors would have certainly included the risk of negative results and wrong conclusions. Nevertheless, in order to determine the value and efficacy we tried to let be carried out the statistically provable clinical tests of the enzyme therapy of cancer at proper university and cancer clinics. This endeavor brought forth only relatively few utilizable informations on account of unexpected difficulties, mainly in German clinics.

At more than 46 university clinics and other large hospitals suitable for testing of cancer therapeutica the investigations of this therapy were introduced, whereby the conditions of double-blind tests were partly included in the plan. In almost all cases the exact evaluation failed for the following reasons:

1. The therapy with proteolytic enzymes can have a lasting therapeutic effect only if it is continued over longer periods. A long-time therapy could not be performed in most of the larger planned investigations. The cancer patients remained only 2 to 8 weeks at the hospital in the surgical or radiological wards. Then they were turned over to their private physician or an after-care sanatorium and came back after recurrences or metastasis, mostly in a far-advanced stage.

As a rule the clinical investigations were, when possible, so organized that the enzyme therapy started already during the surgical or radiological treatment periods. After discharge of the patient, his physician or after-care clinician was advised in a detailed transfer letter, how to continue the treatment. The patient was supposed to return periodically for examination at the clinic where the treatment had started.

An example of it is the blind test series at a university radiation clinic in 1966/67. Irradiated cancer patients received simultaneously Wobe-Mugos* preparation; after discharge from the hospital the physician taking care of the after-treatment was supposed to continue according to the direction of the hospital. At a control test after 6 months, about 60% of the patients replied that they discontinued the therapy either by their own decision or on advice of their physician. It may be presumed with great probability that of the remaining 40% at least one half did the same, so that only 20% of the originally treated patients had

---

* Manufacturer: Mucos, 8022 GRÜNWALD, GERMANY

continued the therapy. Such results are entirely worthless for the evaluation of long-time therapy, especially regarding the examination for metastasis frequency.

Another example of the difficulties connected with an evaluation of long time therapy: At the after-care clinic *Wolf-Zimper,* Bad König at Odenwald, during the last 10 years over 13,000 cancer cases were treated with proteolytic enzymes in combination with the usual treatment methods. Again the physicians in charge of the after-treatments as well as the patients were instructed to continue the treatments according to plan. G. *Stojanow* of the gynecological clinic of the University Varna (Bulgaria) worked over 24 months at the above clinic in order to determine success or value of this therapy. Lists and questionnaires were sent to all patients and private physicians following up their after-care. But only 48% of the physicians and 65% of the patients answered the questions. A locally limited intense inquiry supplied the fact that less than 7% of the patients have been treated consequently after discharge. Therefore, here also a valuation of results of a long-time therapy was out of question. There is no use giving further examples; it was everywhere the same story.

2. According to American investigations, only about 45% of ambulant treated patients are taking their prescribed tablets or suppositories. This is particularly the case with cancer patients who are not informed about the nature of their sickness.

3. On account of the some time observed local pain by the injections and infiltrations where the injection was given, the therapy was discontinued; also among these patients many were not informed about their condition.

4. Frequently a statistical evaluation of clinical investigations is impossible on account of the complexity and heterogeneity of the patient material, furthermore due to the necessity to administer also other therapeutic measures next to the enzyme therapy. Then it cannot be with certainty determined which therapy had been really responsible for success or failure.

Our efforts to work out statistically usable evaluations with the help of a large clinical patient material have been a failure up to date; this forced us to draw upon a multiplicity of individual observations and results among smaller patient groups in order to arrive to a fair, yet critical evaluation of results obtained. This is tried by casuistics to follow later on.

EXPERIENCES AND OPINIONS ABOUT
ENZYME THERAPY IN MALIGNANT TUMORS

During recent years we received numerous reports from West Germany and other countries about the application, experiences and results in malignancies with enzyme therapy suggested by us. These reports are all in agreement with the very good therapeutic results in various forms of cancer and are pointing out again and again that neither disagreeable side effects, nor undesirable late reactions were ever observed.

Especially valuable were the suggestions of our colleagues for the determination of the best guidelines for dosage. There was a uniform agreement that enzyme treatment can only be effective and successful if started *early*, if the dosage is *sufficiently high* and the therapy be *continued* straight along after the apparent subsidence of the active illness.

Several reports show the possibilities of the enzyme therapy in cancer and its limitations. In order to describe clearly this form of therapy, it will be discussed more in detail in the following reports.

Neither operation nor X-ray treatment or a combination therapy leads to a complete recovery in a satisfactory percentage of cases. The tendency of sudden recurrence hangs over the patient for the rest of his life as a continual psychic trauma. Therefore, besides the classical therapy with knife and rays, a treatment is urgently required which has been proved to cause marked damage to the cancer cells, which slows down the progressive growth and counters the formation and spreading of metastases.

These conditions are largely met by the enzyme therapy. In this monograph *J. Beard* has been mentioned repeatedly, he may be considered the founder of the enzyme therapy.

But only recent years have brought the connections between fibrinolysis, thrombolysis and metastasis into focus, by means of animal experiments. Connected with this was an entirely new approach to the interpretation of pathogenesis, clinic and therapy of the cancer disease. The local cancerous tumor, which from the point of view of histopathology is an isolated process, was found to be only a symptom of a general illness which is involving the entire organism. Clearly, therefore, an operation or irradiation of the tumor presents only an isolated local attack which does not take into account general factors and involvement of the rest of the body in this disease. Additional methods of approach are therefore built into the treatment plan, for instance cytostatica, specific diet combinations which help to reduce cancer growth, procedures to enhance the general resistance of the body, and many others.

It cannot be denied that at first cytostatic methods of treatment lead in many cases to surprising successes. But at the same time it appeared very soon that life-threatening side effects prevented their use on an extended basis and thus restricted this therapy only to certain types of tumors and as a palliative measure.

Problems which arise from the use and evaluation of new methods stand today in the center of discussions between clinicians and pharmacologists. Great difficulties usually arise during the evaluation of new medicaments. If the disease type, its location and condition of the patient allows, surgery or radiation will be mostly indicated. But with proper additional methods one must try to influence the growth of the cancer and the formation of metastasis. Final results can be only stated after the patient's observations over a period of years. The regular follow-up examinations requested by clinicians are rarely possible for a variety of reasons (cost, lack of personnel, etc.) yet remains a vital necessity. Furthermore, a closer cooperation between clinic and the physician following up therapy could certainly be improved in many respects.

Therefore, it would be welcomed if as many physicians as possible would communicate their experiences and special methods of treatment to a broad group of their colleagues. In the often short time of observation a suitable complete report for publication is not often available. The patients material covering the treatment with protease mixture Wobe-Mugos is quite extensive and is partly discussed in this monograph. It shows the physician possibilities in specific cases of this therapy and how to follow up the results.

In the course of the past ten years at the clinic of *Wolf-Zimper*, over 13,000 cancer patients received follow-up treatment with proteolytic enzymes. In cooperation with the State Insurance

Companies, this therapy is instituted on a broad basis, whereby the method varies from case to case.

The following table furnishes a review according to the types of tumors of the various groups of patients who were treated in 1963 and 1964 with the enzyme combination Wobe-Mugos:

Carcinoma (in various organs)	1963 Cases	1964 Cases
Mammary	143	63
Ovary	22	4
Uterus	260	111
Vulva	4	1
Male genital tract	15	8
Mouth, tongue, larynx	10	2
Intestines	32	9
Stomach	33	10
Liver, gall bladder and ducts, pancreas	3	4
Hypernephroma	5	3
Bronchi	18	3
Brain tumors	3	1
Skin	9	6
Leukemias-lymphogranulomatosis	30	3
Total	587	228

In 91 patients of 1963 metastasis could be clinically shown. Of 587 patients, 408 were treated with X-rays and radioactive substances before the beginning of our after-treatment.

Of these, 20 patients had metastatic growth. X-ray or radium treatments were given to 158 patients. A short statistics of these years gives the results of the after-treatment with Wobe-Mugos, whereby comparative groups of patients were selected:

1963	587 patients	1964	649 patients
improved	65.1%		68.0%
static	30.9%		21.4%
worse	4.0%		10.6%
able to work	43.0%		39.0%

From these figures it appears that during the years 1963 and 1964 of all patients 65% respectively 68% of the patients improved through treatment with Wobe-Mugos. 25% of the patients could be brought to a passable stationary state of health, while only in 4, respectively 10.6% the condition deteriorated.

From the sociological or economic point of view, it is of great importance that approximately 40% of the patients could resume their former useful activities.

The follow-up radiation treatment causes still so often radiation sickness and, as a late result, considerable permanent indurated scar tissues, also, edematous swellings which are difficult to control therapeutically, especially of the upper and lower extremities. With various substances, such as vitamins and tissue extracts, attempts were made to reduce such side effects of radiation therapeutically. However, none of these methods showed much success.

Wolf-Zimper shows in his large number of clinical cases that patients take the after-radiation therapy with little side effects, best, if the enzyme combination is given before, during and after the ray-treatment. Symptoms which can be observed objectively, such as leucopenia, thrombopenia, erythema as an expression of local inflammations, as well as the subjective symptoms of radiation sickness in the form of nausea, diarrhea and severe general sick feeling appeared only in a very reduced amount or not at all. The completion of the radiation treatment was possible without disturbance, since the patient was hardly aware of these radiation side effects.

Also the late symptoms of an intensive radiation treatment, such as radiation fibrosis, radiation ulcus, keloid formation and extensive lymphedema are fully or greatly avoided by the use of the enzymes. In this connection it may be mentioned that in several radiation clinics the combination treatment of radiation and Wobe-Mugos is already being introduced as a matter of routine.

Stojanow reports about the use of the enzyme mixture with 974 female patients suffering from tumors in the genital region. Most successful proved to be enzyme clysmas (retention enemas) which enables the physician to administer even very large doses over long periods without complications.

Remarkable successes are seen after intratumoral injections.

After injections (in some cases under anesthesia given), mostly already after the second or third treatment a liquefication of the tumor ensues which can be drained off by punctation. In many cases it is possible to restore these gynecological tumors by the enzyme mixture so far that any urgently required operation can be done. Partial dissolution of the tumor or regression to a great extent (especially with carcinomas of the ovaries) are often seen. With incurable patients the growth of the tumors is significantly slowed up till the lethal end takes place less dramatically.

A review of his experiences on enzyme therapy was given by *Titscher* (Chief of Thorax Surg. Dept., Hospital Lainz). He demonstrated 38 cases of primary pleurocarcinosis and pleuritis exudation due to metastasis of inauria carcinoma, to whom he injected into the pleural cavity 1000 mg Wobe-Mugos every 2nd or 3rd day. In 34 cases he observed total remissions. The prognoses and general conditions were definitely improved, the patients were free of pain and respiratory handicaps.

Wrba pointed out that these results give one the conviction of a definite success and this therapy should be used in all patients with such diseases. After enzyme therapy, many patients could successfully be treated with other therapeutic forms, e.g. surgery, which was not possible before enzyme treatment.

Hoefer-Janker et al. gave Wobe-Mugos intratumorally or into cavities in high doses. They were able to liquefy tumors after repeated injections of 10-15 ampules of this preparation into tumors which already have been irradiated to the limits of tolerance. Tests showed that the liquid drained off such tumors was negative regarding tumor cells. In many cases all that remained of a tumor treated this way was hard scar tissue.

The first female patient treated by this method who had an adenocarcinoma the size of a child's head of the corpus uteri, is still alive after more than 2½ years and is free of tumor or other complications.

As a rule, all tumors and their metastases which can be reached by injection may be treated with Wobe-Mugos injections. 40% of all tumors treated this way will show a positive reaction or regression. Particularly responsive are mamma- and ovarian-carcinomas and their metastases. Conglomerates of tumors or neoplasms of the gastro-intestinal tract should be treated similarly.

Encouraged by these results of intratumoral injection, *Hoefer-Janker et al.*, during the past year have again in greater measure injected the enzyme mixture into the cavities of carcinoma of the pleura with exsudations.

In this connection the informations of *Daum* were of special interest. He discussed the principal questions of dosage and application and furnished impressive examples from his clinic.

Daum used wherever possible the intratumoral application of the enzyme mixture which has been especially successful with large, inoperable ovarian carcinomas. Injection into the parametrium permits high dosages and sufficient enzyme activity in the tumor. Under narcosis several ampules Wobe-Mugos are injected intratumorally through the abdominal wall. The dose depends on the size of the tumor and the general condition of the patient. In order to obtain the highest possible initial level of the enzyme mixture within the tumor and the surrounding tissue, generally 8 to 10 ampules are given at once. Radiation treatment may follow later by giving the standard dosage. Clinical investigation proved that the tumor is more "radiation-sensitive" under enzyme treatment and the shrinking process of the tumor is accelerated. Radiation was tolerated much better after or during enzyme treatment (*Barth, Keim*). The patients show no lymphedema and no or markedly reduced vomiting.

Dorrer used the enzyme therapy on more than 100 patients and could ascertain that the tumor growth receded in most cases quite visibly. The life expectancy was appreciable extended, recurrences were prevented. Also several patients who became resistant against the previous therapy (e.g. chemotherapy), improved promptly upon the enzyme therapy. *Dorrer* varied the dosage and type of application depending upon the cases. He often applied three ampules of Wobe-Mugos on three days a week and gave additionally 12 tablets or one to two suppositories daily. Under this treatment the tumors were mostly reduced in size or even disappeared and the infiltrative growth was inhibited.

Some inoperable patients could thus be brought to a condition where an operation became feasible. It was noteworthy that not a single metastasis ensued, as long as the patients were held on the enzyme therapy. The euphoria which could be observed among his moribund patients during the enzyme treatment was a desirable and welcome effect. The exitus if at all followed suddenly and undramatically without any agony.

Dorrer's success was apparently primarily due to the consistent administration of the enzyme treatment. The therapy was not discontinued too early and the physicians in charge of the after-care, as well as the patients themselves, were emphatically advised that the enzyme treatment would have to be followed up for a long time, often for many years or a life time, in order to avoid recurrences.

To avoid repeated injections, the regular dosage of daily 1 to 4 suppositories was later recommended, which proved successful.

In order to exchange experiences, 76 leading physicians and specialists met during the second "Enzymology meeting" in Nuremberg (1966).

All participants stressed the point that the therapy with enzymes, in contrast to cytostatic treatments, had no side effects whatsoever, there is no danger of overdose.

The main problem in certain tumor types is that since they are mostly very poorly vascularized, some proper techniques have to be found to introduce as large amounts as possible of the enzyme mixture into the malignant tumors.

All participants at the convention agreed emphatically that under enzyme therapy the number of large post-operative hemorrhages diminished and that potent analgesics could be avoided. A light and very desirable euphoria eases the final stage for those patients whose condition is too far advanced for effective improvement. In order to avoid recurrences after the initial therapy, an ambulatory treatment should be carried out by all means over a longer period after discharge from the hospital, in cooperation with the family physician. The administration of the enzyme combination in form of suppositories or retention enema tablets or lozenges, has proved very useful in the prophylaxis of metastasis. In cases of operable and inoperable pulmonary carcinomas the enzyme treatment usually leads to a tumor regression and a considerable improvement of the symptoms and the general condition.

In pleuritis carcinomatosa the fluid level is reduced and the dyspnea diminished. The intrapleural injection of the enzymes should be tried, so that an adequate enzyme level is achieved in the involved pleural cavity (*Herzer*).

Lohmüller points out in his contribution that urologists decline the use of radiation treatment, since no success could be observed. In contrast to other specialists, urologists are often in a position to check up the extent and changes of the tumor cystoscopically.

In five patients, including two over eighty, with inoperable bladder carcinomas the results after enzyme treatments were surprisingly good. In all 5 patients tumor remissions took place. Since the bladder carcinoma is generally well vascularized, the enzymes reach the tumor presumably via the blood stream and act there, like a local injection.

Plazer-Altenburg reviews 42 cases of prognostically bad carcinoma cases which reacted most favorably to the enzyme treatment. He stresses the point in his conclusions about the result of his enzyme treatments that if treatment is started early enough a definite stop in the metastasis formation through Wobe-Mugos is possible and can be proved. He also prefers the intratumoral injection, if it is anatomically feasible.

Freihofer dispensed Wobe-Mugos to over 250 patients, mostly operated or radiation-treated cases (malignant cases of mammary gland, stomach, colon, bronchi, of bones, thyroid and one hypernephroma).

His investigations proved that systemic Wobe-Mugos in form of i.m. injections either considerably retards malignant growth or can stop it completely. Often a reduction of the size of the tumors was observed, a complete dissolving, however, did not take place in any of the systemic treated cases, which could hardly be expected in his extensive tumor cases, especially since the larger part of the patients were already in the terminal stage.

After intratumoral injections large dissolved tumor areas could be observed. Blood picture controls showed no adverse influence by the enzyme combination; undesirable side effects were likewise not in evidence.

The third symposium on enzyme therapy took place on November 31, 1968 in Nuremberg.

56 heads of private clinics and department heads of university clinics and major hospitals reported their experiences with enzyme treatments of malignant tumors of various types and locations. They showed impressively and clearly the therapeutic possibilities of enzymes, but they brought out just as clearly the problems involved and the many unsolved questions.

Schneider offered an impressive number of interesting successful case histories. An inoperable ovarian carcinoma with peritoneal involvement became soon operable under enzyme therapy. He saw surprising improvements in cases of stomach and colon Ca. While a number of them finally succumbed, marked regressions were seen in their tumors, also those of the urinary bladder and mamma. His observation on his cancer cases convinced him that enzyme therapy influences favorably the course of most cancer diseases, without any drawbacks in using even massive doses. He used minimum doses of 300 mg daily, in some cases a multiple of it. His casuistic included many cases with objective tumor remissions due to enzyme therapy.

Werkmeister described in detail 2 cases of reticulo-sarcomatosis cutis successfully cured by large doses of Wobe-Mugos, with necrosis of the tumor and extrusion from the wound, also reticuloendotheliosis controlled by it. A lip carcinoma was cured by intratumoral injection.

Klose stressed the point that in the treatment of inoperable cervical malignancies, no long-term therapy can be given with any remedy except the enzyme therapy and his success with internal therapy of advanced cases are only built upon enzyme therapy. He demonstrated a number of successfully treated patients.

Schnellen saw excellent results in the treatments of malignant melanoma in 14 cases with enzyme injections and suppositories. Small tumors disappeared and did not return. Moreover, large tumors could be dependently eliminated by this therapy alone.

The group of scientists and physicians agreed that for the long-term therapy, especially for metastasis and recurrence prophylaxis, a dosage of 300 to 400 mg Wobe-Mugos easily proved desirable. At present no final decision can be made whether a continuous application or intermittently higher dosages are more appropriate. They give some of their patients a daily dosage of

600 to 800 mg Wobe-Mugos over a period of a month and then interrupts it for one month. They think to realize better results thereby and reason that a substantial part of the administered enzymes becomes inactivated by inhibitors and that only the enzyme quantity becomes effective which remains after this inactivation.

For the prevention of metastasis, a considerable increase of fibrinolysis must be induced at least periodically so that the colonization and proliferation of metastatic cells is stopped, respectively micro-aggregates of tumor cells already established are dissolved or eliminated. From this point of view the intermittent high dosage makes sense.

It can be taken for granted that patients react more favorably in conjunction with a sensible diet. It is important that animal fats, excepting small quantities of butter, are excluded completely and the carbohydrate consumption is reduced to a minimum, as sweets.

Optimal therapeutic results are doubtless achieved by local or topical applications with Wobe-Mugos. This applies as well for the external treatment with Wobe-Mugos ointment as for the intra- and peritumoral injections and also for the intrapleural and intraperitoneal infusions. With intrapleural applications, once the puncture and drainage of the pleural exsudate is accomplished, one to four ampules of Wobe-Mugos are instilled, dissolved in approximately 10 ml. saline.

The intraperitoneal application occurs similarly with 4 to 10 ampules Wobe-Mugos dissolved in 30 cc saline solution administered through the ascites-puncture.

The instillation into the bladder of high dosages of Wobe-Mugos in urinary bladder carcinomas and by retention enemas in rectal and colon carcinomas has proven very successful. With bladder instillations approximately 10 ampules of Wobe-Mugos should be applied as a watery solution, for rectal instillations a dose of 1000 mg minimum.

While Wobe-Mugos is free of any side effects, especially upon the hematopoietic system, after intramuscular applications at times painful reactions and indurations follow at the injection site, which clear up mostly within a short period. Now and then, but very rarely, allergic reactions during injection series are being observed. In these cases we must differentiate between true allergies and circulatory "overload", with similar symptoms. The enzyme treatment leads to production of tumor breakdown products, mainly from the border zones of the tumor. These products may temporarily over-burden the blood circulation and liver to some degree and thus cause symptoms which resemble an anaphylactic reaction.

In true allergies, as *Schnellen* pointed out, there is a difference between the very rare enzyme allergy and an autoallergy in which the protein breakdown products of the tumor act as allergens. These allergic reactions are mostly weaker and much rarer than the penicillin allergies observed so frequently nowadays. Since allergic aspects were observed up to now only after parenteral enzyme use, there is almost always an alternative chance for rectal and oral application.

Westrick outlined his statistics on successful prevention of metastases by enzyme therapy. He pointed out the connection between fibrinolysis and metastasis formation. During the International Symposium for Coagulation Research held in Vienna, noted coagulation researchers reported unanimously that a significant reduction of the metastasis formation was obtained through fibrinolysis induced by medication.

In order to develop any metastasis, cancer cells circulating in the blood or through the lymphatic ducts must be surrounded by a fibrin coat, by means of which they can stick to the vessel endothelium. If it is possible to change the viscosity or fibrin stickiness which conditions this adhesiveness, the forming of metastasis is prevented.

Based on many animal experiments this metastasis-prevention effect of fibrinolytic agents could be definitely demonstrated. In Wobe-Mugos is a fibrinolytic agent which performs this task in an ideal manner, for it is suitable, on the one hand, for the long-term therapy; on the other hand it contains additional selective cytolytic properties specific to tumor cells. In contrast to a cytostatic therapy, we have here a possibility whose advantages are obvious and which makes it especially suitable for metastasis prophylaxis.

Lisicky (Czechoslovakia) pursues since January, 1967 systematic investigations with simultaneous controls of the laboratory figures in 52 cancer cases. His casuistic contained many patients with lethal prognosis and represents therefore a definite negative selection.

The very carefully undertaken follow-up examinations of the patients showed also here beyond doubt that under enzyme therapy metastasis formation is greatly reduced or absent, that the general condition of the patient improves and the use of potent pain remedies, like opiates, could be reduced or stopped. The clinical evaluation of the findings showed that mammary and rectal carcinomas react best to the enzyme treatment. Also, with inoperable tumor patients the survival time did not become a period of suffering, whereby the impression often arose, that the final stage was considerably extended and much better bearable.

Lisicky points out that the enzyme treatment is at least equal to, if not superior to any other known additional cancer therapy. The laboratory findings indicated no abnormal values. The compatibility was excellent. After intratumoral injections of the enzyme mixture the rarely appearing allergies were short and caused by the sudden dissolution of tumor cells.

Also *Rosanova* (34) paid particular attention to the enzyme therapy of cancer patients. From his large number of cases he discusses some especially characteristic cases which indicate a clear influence upon the metastatic processes.

Female, 52 years of age, radically operated mammary carcinoma with axillary metastases. After a few years a papillary metastasis appears in an appendix scar, also a metastasis in the 2nd L.V., after 2 months also in the 5th L.V. The lumbar vertebrae were destroyed. Further metastases in the right lung and in the ileum. The enzyme treatment leads already within a week to a reduction of pain. In the second week all tumors are regressing. Four weeks later X-ray indicates no visible tumors. Aside from a shortening of the right leg, the re-examination shows no indication for any malignant processes.

Mammary carcinoma after surgery and irradiation within 3 months extensive metastases in the lungs. The enzyme treatment brings about complete recovery of the patient, she is able to work again. X-rays after 2 years shows an almost normal pulmonary picture.

Bartsch and *Gosemärker* in their publication reported that 21 post-surgical patients (Stage II) who had received intensive Wobe-Mugos prophylaxis for 2 to 3 years, all remained free of metastasis. In another group, in 3 of 50 patients with metastases objective and distinct complete tumor remissions took place, while with 20 patients the illness took a less progredient course. All these cases represented far advanced stages with lethal prognosis, advanced metastasis and pronounced tumor cachexia. A result of their low condition was the extremely reduced capacity of absorption, even very high enzyme doses in some cases (40-100 g) did not influence any more the disease. The author is of the opinion that prolongation of life can mostly be obtained with the enzyme treatment, but this is naturally very difficult to prove statistically.

Similar to many other authors, also *Bartsch and co-workers* point out in their treatise that an early beginning of the prophylactic enzyme therapy is decisive for the success, assuming the dosage is sufficiently high and that the treatment is not discontinued prematurely.

According to their experiences, the prophylactic administration of enzymes for prevention of metastasis is even more promising than the therapy of metastasis already formed. Just as it has become accepted to undertake major operations under the protection of antibiotics, enzymes should be given prophylactically before, during and after tumor surgery, also during irradiation. *Bartsch* published his very positive therapeutic results in 71 advanced cancer patients whereby in all of them enzyme therapy in high doses was continued a minimum of 3 years.

Impressive also are the observations of *Winkler*. With a juvenile patient with a retothelial sarcoma total disappearance of the tumor followed the administration of the enzyme therapy.

In another patient with an operated, metastasizing pancreas carcinoma, the blood picture and the general condition improved markedly with the use of Wobe-Mugos. Also 4 cases of bronchial carcinomas and a case of primary carcinoma of the gall duct reacted very favorably to the enzyme treatment. The conditions remained stationary for several months.

Horojshi writes also that after enzyme treatment, patients with metastasizing gastric carcinomas became free of pain. In contrast to other authors, he saw no success in the enzyme treatment of plasmacytomas.

In order to achieve some therapeutic results in cases with lethal prognosis, the enzyme treatment must begin as early as possible, as mentioned repeatedly before, and the dosage selected must be as high as possible. These connections are seen by the following case:

A patient with a chondrosarcoma of the right hip bone was not operable, due to intra-abdominal metastases. The therapy was started with Wobe-Mugos (tablets and suppositories in high doses). To hope for a basic improvement or inhibition of the cancer growth was certainly out of the question. With the aid of the enzyme treatment it was possible, however, to sustain her for 9 months relatively free of complaints.

Also the nose and throat specialists supplied reports about the use of the enzyme preparation in malignant cases.

Desa (India) cured a number of neoplasms of larynx and nose, *E. Maier* squamous epitheliomas of tongue and larynx, also leutoplakias promptly disappeared by enzyme therapy.

Sanz-Anton pointed out in his lecture at a Madrid Symposium about cancer chemo-therapy that cancer patients tolerate radiation treament much better, if Wobe-Mugos is used at the same time. The radiation sensitivity of the cancer tissue as well as the radiation tolerance is enhanced.

Biochemical tests have shown that the extracellular oxygen concentration rises while the intracellular level is reduced. All of the 25 patients tested in this respect (uterus, cervix, vulva and thyroid carcinoma, as well as a lymphatic sarcoma) were considerably improved through the enzyme protection after conclusion of radiation series. Especially their food intake, physical strength and body weight increased very much, anorexia disappeared. At the clinic of *Sanz-Anton* the combination treatment—enzymes plus radiation—is now used on a large scale since this proved superior to all other combinations as indicated by the positive results of therapy.

Emeterio (Spain) underlines also that a long interruption of the enzyme treatment should be avoided under all circumstances. His investigations show clearly that the treatment must be continued until the tumor has completely disappeared, or until a long lasting standstill of the tumor growth has been established. His casuistic contains several cases of bronchio-genic carcinomas with objective tumor remissions.

The following form of treatment has been successful at the Clinic of *Lopez de Osa:* Large tumors are removed surgically as radically as possible, followed immediately with large doses of the enzyme. Already in the first month of treatment it can be proved that the metastases are visibly reduced. Altogether 19 patients with gynecological tumors, all of them resisting any other forms of therapy (operation and rays), were subjected to enzyme treatment. In all cases the disease was far advanced and the prognosis was hopeless. Considerable improvement was achieved in 6 cases, while a definite cure could in no case be expected. The author found that the enzyme therapy supports radiation treatment very effectively and exceeds all other methods of treatment, also by its outstanding tolerability.

After all these detailed and impressive numbers of cases in different countries presented and physicians, specialists and clinical observers of different branches of medicine related their findings and opinions, I, *Wolf*, would like to report some of my friends' and my own experiences, collected during the last 23 years.

During this period our group treated 1139 cancer patients, of whom we exclude 193 cases from a critical evaluation. These were either less than two weeks under our care, did not continue for unknown reasons, or died in the interim from other causes than by the disease itself (such as accidents) or due to special other reasons.

945 of these cancer patients had histologically confirmed diagnosis. However, an exact division into tumor stages is hardly possible because operations, or other therapies had preceded and the initial diagnosis as to the grade and extent of the illness could hardly be ascertained any more in all cases.

The table furnishes charted information about the composition of the casuistics and some details of the treatment. It shows that more than 60% were treated surgically and only approximately 40% were left without operation due to particular conditions of the tumor or the patients. Only approximately 10% of the patients received exclusively enzyme therapy, of whom about 2/3 suffered from skin carcinomas.

Types of cancers till the end of 1967 and therapies used.

Type of tumor	Numbers	Operated	Not Operated	Treated exclusively with Enzymes	Combined with other treatments
Mamma	261	207°	54	9	252
Ovaries°	49	42*	7	4	45
Corpus	41	32°	9	0	45
Cervix	62	24	38	0	62
Vagina	2	2	0	2	0
Ur. bladder	16	11°	5	6	10
Prostate	49	11	38	0	49**
Adrenals	8	1	7	0	8
Esophagus	3	2	1	0	3
Stomach	31	17*	14	0	31
Colon	45	35*	10	0	45
Rectum	71	44*	27	0	71
Liver	2	0	2	2	0

Pancreas	9	1	8	0	9
Lung	38	11	27	4	34
Brains	6	6*	0	0	6
Throat-Nose-Ear	17	14	3	0	17
Skin	98	31	65	52	44
Malig. Melanomas	9	3	6	0	9
Reticulosis					
Hodgkins	42	0	—	—	42
Others	18	6	12	0	18
Chronic Leukemia					
Lymphatic	14	0	—	—	14
Myelogenous "	9	—	—	5	4
Others "	6	2	4	0	6
Sarcoma	38	11*	27	0	38
	944	513	364	84	862

* Partly, only partial tumor resection
** Hormones

All other additional measures employed in the cases treated on a combined basis, included cytostatica, hormones, vitamins (primarily A, E and C), roborants, heparin, diet and general health items.

Until 1959 we used singly, or in combination, the following enzymes for the therapy: Trypsin, chymotrypsin, papain, ficin, bromelin, desoxyribonuclease, fungal proteases as well as lipases, and extracts from calves' thymus and of leguminoses. Since 1959 we used exclusively Wobe-Mugos in its published forms of preparation and constituents. After employing dosages of 50-100 mg i.m. or 200 mg orally until 1964, we now found that better results can be achieved with a dosage of 600 mg or more per day. Since 1969 we use systemic doses of 2000-4000 mg enzymes per day (see directions for enzyme application) to the tumor.

We had the best therapeutic results with direct application of the enzymes. This is true for Wobe-Mugos ointment with skin tumors as well as with intratumoral and peritumoral injections or as intraperitoneal and intrapleural or bladder infusion.

In ascites formation, especially in ovarian carcinomas, we give 6 to 10 ampules Wobe-Mugos 1 to 3 times per week each time after aspiration of the ascites, through the same needle, in pleural exudates each time 2 to 4 ampules, also after the aspiration. These forms of administration have proved so successful that we intend to considerably intensify their application. With bladder tumors we use catheter instillations of 2 to 10 ampules each, up to now also with satisfactory results.

It is noteworthy that with the injections just mentioned as well as in the infusions and the bladder instillations hardly any allergic reactions occurred. In a few cases in which local pain was caused by intramuscular injections, we furnished daily 1 to 4 ampules Wobe-Mugos in 200 cc drop infusions during 2 to 4 hours. In these cases we had occasional allergic reactions, urticaria and high fever, but never life-endangering symptoms.

Our instructions to the patients to continue taking the enzymes after the intense therapy was seldom followed. After some time of well-being the enzyme intake was broken off or interrupted. 2 to 3 months later in many cases a renewed deterioration began with often rapid tumor proliferations.

The following discussion of tumor cases in individual organs or organ systems shows clearly the similarity in the results, just as repeatedly confirmed, also by other therapists, and as it will certainly be observed by others if the therapy is followed up consistently and for a long time.

Mammary Carcinoma

Mammary carcinomas and their metastases respond well to proteolytic enzymes. Of 207 operated cases with primary mammary carcinomas we could put 122 under an intensive enzyme therapy, after consulting with their operating physicians, already before and immediately after the mastectomy.

The treatments were continued after the hospital discharge as long as possible on account of metastasis prophylaxis.

Insofar as it was not absolutely required on account of the tumor situation, we did not advise a radical lymph gland resection in the axilla, also no pre- and since 1964—no post-radiation treatment. We believe that the enzyme therapy at least replaces successfully these measures and even considerably surpasses them as far as metastasis and recurrence rates are concerned.

Of 107 statistically evaluable female mammary carcinoma patients of different stages with mastectomy, after 5 years 90 still survived; that is, 84%. This result is considerably better than the average of other statistics, which likewise disregard dividing their cases in TNM-stages (Germany 1957 48%, Mayo Clinic 43%).

Here we must consider that a larger number of these patients interrupted the enzyme therapy after their mastectomy for a long period of time. This was apparently the cause for the lethal outcome of a whole series of cases. With enzyme treatments which have been continued persistently over a 3 to 5 year period, a considerable increase in the 5-year survival rate can certainly be expected.

A large part of the patients came to us only after recurrence or metastasis had already developed after the mastectomy. As far as possible, metastases of the skin, the axilla or in the scar tissues were removed and at the same time Wobe-Mugos was given in large doses intra- and peritumorally as well as systemically.

An exact recording of the 5-year survival figures of these patients is difficult, since many did not return to us for further treatment. In 52 known cases with operable skin and axillary metastases, 29 survived the 5 year limit. This result, too, exceeds comparative statistical figures.

Reductions in size and total disappearance, mainly of smaller tumors, were observed very frequently.

In extensive ulcerating tumors of the skin we applied, besides the systemic enzymes, several times daily about 1 mm thick layers of Wobe-Mugos ointment over the ulcerated areas and injected wherever possible at the same time intratumorally. Approximately 2/3 of the cases cleared up, the ulceration partially closing, the tumors reducing to about 50% of their sizes and stabilizing the tumor growth up to 5 years, especially in elderly patients. It is remarkable that these patients retained a general feeling of well-being for many years in spite of the large, ulcerating tumors, they performed their housework in many cases.

Pulmonary metastasis was remarkably rare, as long as the enzyme therapy was continued. In 26 female patients under enzyme therapy with stationary skin metastases, in most cases for several years, within 2 to 6 months after the interruption of the treatment lung metastases and at the same time tumor-cell positive pleural exudations occurred.

As expected, the results in treatments of metastasis in the lung were less convincing. Next to the usual therapy we applied high dosages of Wobe-Mugos. In case of pleural effusions we injected after aspiration 2 to 10 ampules of the enzyme combination 2 or 3 times weekly. In almost all cases the exudates regressed within approximately two weeks, they became tumor cell negative and clear, breathing difficulties and cough disappeared, and general symptoms improved to the point that the patients became hopeful again. However, after periods of improvement between 3 and 30 months, a renewed deterioration of the general condition set in with the well known consequences. Remarkable was the very short final stage, which took a much more tolerable course than experienced in patients treated differently; this we consider with some justification as a certain success. A moderate number survived 5 years.

Tumors of the Female Genital Tract

The results of other researchers who report about malignancies in areas of the female genital tract agree fully with our experiences. The proteolytic enzymes have proven themselves well in this field. In suitable cases the enzyme treatment was combined with surgery or radiation.

Ovarian Carcinoma

Impressive are the results after enzyme use in far advanced inoperable ovarial carcinomas with concomitant ascites development. We gave Wobe-Mugos in high doses orally, rectally and i.m. and could achieve improvements of the severe clinical picture in 7 out of 12 cases. Large, palpable tumors became softer, the amount of ascites diminished and small metastases disappeared at times completely. The patients showed a significant lift in their general well-being, with increase of appetite and weight and decrease of pain. The remissions lasted 2 to 28 months, at the same time the tumor growth stopped. Radiation and enzyme therapy combined led to better results than the rays alone.

Since 1965 we gave Wobe-Mugos locally, as *Daum* applies it in his clinic. The content of 6 to 10 ampules Wobe-Mugos was applied intratumorally 2 to 3 times weekly. Frequently we noted already after two to four of these treatments a liquefying of the tumor, so that a considerable part of the fluidified cancer tissue could be drained off. After applying 8 to 10 ampules intraperitoneally, the ascites volume was reduced or it disappeared altogether, and the abdominal tension regressed.

Of altogether 11 of these patients receiving this treatment more than 2 years ago, nine are still alive. However, of these 8 were operated on in the meantime or were irradiated, so that the results must probably be ascribed to the combined treatment.

It would be desirable to ascertain statistically in larger clinics the success of this type of treatment, because with our few cases not only the objective and subjective results of the treatment were remarkable but also the survival span: all of the 11 patients with ovarial carcinomas of stage III and IV showed a survival time of over 20 months. This number at any rate is well above comparable values.

Corpus Carcinoma

The success of the treatment of corpus carcinoma cases was in general above average. All 41 patients had received prior to or parallel to the enzyme treatment combined radium and X-ray therapy; 32 were operated. Noteworthy was the good tolerance of the radium and X-ray therapy with simultaneous enzyme administration. The reduced rate of recurrences and metastases after radiation or operation of patients in Stage I and II, we believe, can be traced directly to the enzyme treatment, but not sufficient figures are available for statistical purposes.

In cases of corpus carcinomas Stage III and IV, as a rule the improvement of the general symptoms was remarkable. The life span was apparently extended several months, at time up to a year, the final period was short and bearable, analgesics were hardly required.

Collum Carcinomas

According to our observations, Collum carcinomas respond much more favorably to proteolytic enzymes than corpus carcinomas. 51 of our 62 patients came to us only after preceding radium and X-ray treatments or surgery. 42 of these were free of tumor in the beginning, so that the enzyme therapy was only used as prophylaxis against recurrence and metastasis. Of 19 evaluated cases of this group (Stage I and II) 16 lived beyond the 5 year limit which can well be credited to the prophylactic effect of the enzymes. Among the more advanced metastasized cases a considerable subjective improvement, gain in weight and retardation of the tumor growth could be observed. We often had the impression that more or less prolonged life span resulted with mostly satisfactory general condition. As long as the enzymes were given during or immediately after the X-ray or radiation treatment, X-ray side effects, especially late damages, were noticeably reduced. In none of these cases did we observe phlebitis or thromboses.

Portio-Erosions

In approximately 30 female patients with inflammations or erosions of the portio we gave Wobe-Mugos suppositories vaginally, besides systemic doses. Thereupon in over 90% of the cases the inflammation disappeared in a few days. We are convinced that with the diagnosis of portio erosion, an intensive, local enzyme treatment is indicated, in conjunction with cytological examination.

Bladder and Prostate Carcinomas

In collaboration with specialists for urology, we administered enzymes in confirmed bladder carcinomas to 16 patients, to 7 with bladder papillomas, where the suspicion of carcinoma existed, and to 49 with prostate carcinomas.

In bladder carcinomas, next to the intensive systemic treatment mainly with suppositories, we applied occasionally bladder instillations with 4 to 10 ampules of Wobe-Mugos. In 9 bladder carcinomas and 5 papilloma cases reductions or disappearance of tumors could be proven by cystoscopy after enzyme therapy alone. 8 of these patients however were later operated so that it is difficult to decide to what extent the permanent success could be ascribed to the enzymes. In 5 far advanced bladder carcinoma cases the enzyme treatment brought distinct improvements of the general condition and micturition.

In prostate carcinomas it was difficult to determine an objective influence on the primary tumor with enzyme therapy, apart from some definite subjective improvements, because all patients were also treated with hormones. However, in 16 of 28 patients bone metastasis distinctly regressed, in 5 patients they disappeared completely, which normally is not to be expected with hormone treatment alone. 60% of the patients, who were treated by us for a period longer than 3 years, we have lost sight of so that a statistical evaluation of the survival rates again cannot be made.

Hypernephroma

The enzyme therapy of the few advanced and inoperable hypernephroma cases treated by us slowed down the tumor growth. 4 remained stationary for 3 to 6 months, 2 could be operated later on and painless new growth started 6 and 8 months later. In 2 cases a marked reduction of the sizes of the hugh tumors was produced by intratumoral injections.

Gastric Carcinomas

We treated 31 patients in the course of these years. Wobe-Mugos was given systemically as lozenges and suppositories, also as intraperitoneal infusions. 24 of the cases had come to us in an already inoperable condition. We were impressed also here by the more subjective results with the enzymatic treatment. This therapy, however, can hardly prevent the lethal outcome of an advanced gastric carcinoma but can sometimes achieve perhaps a prolongation of life, up to two years. No generally valid statement can also be offered about metastasis-prophylaxis in gastric cancers.

Colon and Rectum Carcinomas

A more positive aspect is presented in colon and rectum carcinomas as far as the therapeutic value of enzymes is concerned, which we could observe in altogether 126 cases. Of these 79 had been operated once or more. Most came to us only after surgery. 72 of our patients proved to be inoperable, because the primary tumor was too extensive, or because recurrence and metastasis had already set in. Almost all were in a far advanced stage. This negative selection makes the evaluation of a therapy problematical.

In cases of colostomy (anus praeter) we deposited daily 2 to 4 suppositories Wobe-Mugos in the natural anus, and additionally injected 2 ampules i.m. Since 1969 also retention enemas, consisting of 1000-4000 mg Wobe-Mugos enzymes, were repeatedly applied in the artificial exit. Very good local results appeared in the form of tumor shrinking and elimination of even larger necrotic tumor parts. Almost regularly improvement of the general condition, increase in appetite, weight increase and pain reduction were observed. Marked reduction or complete withdrawal of analgesics were possible, diminished micturition complaints and incontinence were signs of improvement. Surprising also was the excellent general condition of the patients in stages III and IV, lasting at times over 3 to 4 years, a result which is not often observed with other forms of treatment.

Although we did not achieve complete recoveries of most cases —and this could not be expected—we think we achieved considerable prolongation of life by the enzyme therapy. In good general condition, which lasted almost always up to the end, more than 70% survived for 3 months or longer and more than 40% of the patients from 6 months to 3 years.

Lucky is the group of 49 patients who were treated with enzymes for the purpose of recurrence- and metastasis-prophylaxis after a successful resection of colon or rectum carcinomas. The beginning of the treatment of 24 cases goes so far back that meanwhile 19 have passed the 5 year lifespan line. In 4 patients recurrences arose; we were able to establish that enzyme treatments had been discontinued over 6 months, and in some cases even 3 and 4 years before. World literature mentions a recurrence and metastasis rate which fluctuates between 30 and 70%, so that we are of the firm opinion that with enzyme therapy carried through consistently, significant improvement of the chances of such patients can be achieved.

Pancreas Carcinomas

Here we were not able to establish unequivocal success of the treatment. Besides, there were only 9 cases, but the improvement in the general symptoms was anyhow distinct with 7 patients. In 5 cases with impending occlusive icterus the complaints disappeared temporarily. But we could hardly assume, and it cannot be proved, that decisive regressions of the tumor have been achieved.

Bronchial and Lung Carcinomas

Almost all patients who came to us with pulmonary or bronchial carcinomas were either primarily inoperable or developed metastasis after successful surgery. They represent again a pronounced negative selection.

We gave systemically high enzyme doses, in pleural exudates also locally, as it was already described in pulmonary metastasis after mammary carcinomas. The improvement of the general condition was clearly noticeable and set in a few days after beginning of the treatment. We observed in almost every case increase of appetite and weight, reduction of cough and haemophthisis, reduction of cough sedatives, also diminishing or drying up of the exsudates, and increased pulmonary load capacity. In 21 of 34 patients the tumor shadow reduced in size in the X-ray pictures, although it is not clear whether this occurred due to tumor reduction or through reduction of the inflammatory zone surrounding the tumor. With 6 patients smaller tumors could not be found any more after 2 to 8 weeks of intensive enzyme application, when compared with earlier X-ray plates.

In more than 50% of the patients a clear stop of the tumor growth set in which lasted for at least 1 year among 22 patients who had been in our care long enough. These were mainly older patients whose tumor growth can be counted on to be less active. We also noted again and again in bronchial carcinomas that after long periods of improvement and stoppage of tumor growth it starts a renewed, progredient increase again if the enzyme therapy was interrupted.

Although we could observe impressive objective and subjective improvements in inoperable bronchial carcinomas, we do not believe that according to experiences gained up to now the enzyme therapy alone can prevent in many cases the fatal outcome even though it may possibly be postponed from 3 months to 3 years.

Especially noteworthy is the favorable general condition which lasts until shortly before exitus. Of 17 patients who died, 9 had been able to leave their bed up to 3 days before death, 2 died while taking walks.

Tumors of Mouth, Nose, Ear, Throat and Esophagus

Almost all patients with tumors in this area asked for the enzyme therapy as ultima ratio when other treatments failed. Yet in most cases the treatment was able to control in a short time ulcerations, bleeding and pain and to slow down considerably the progress. The therapy consisted in the hourly use of 2 Wobenzyme "candies" and 2 suppositories daily. Whenever possible the tumors were treated by intratumoral injections which mostly reduced their size visibly, some disappeared completely.

The reduction of esophageal obstructions relieved the dysphagia. The combination with radiation therapy showed impressive results. 2 tonsillar carcinomas showed remission up to now and leucoplakias usually disappeared.

Skin

Perhaps the most impressive results of the enzyme therapy were seen in skin tumors. It is known that the therapy of basal cell carcinomas of the skin by surgery or radiation is mostly successful. The same successes are accomplished by combined local and systemic enzyme therapy. Of the 64 basal cell carcinomas (ulcus rodens) 90% were cured within 2 to 19 weeks. Epitheliomas and other invasive skin tumors stopped ingrowing or regressed after about 2 weeks under enzyme therapy. In most cases this was followed by a slow, gradual healing, particularly when combined with intratumoral injections.

For precanceroses, like senile warts, often the application of enzyme salve for only a few weeks was sufficient. Two cases of mycosis fungoides reacted with prompt healing of the tumors.

Melanoma

All 9 melanoma cases showed under enzyme therapy very soon changes, most of the small skin tumors reduced in size or disappeared. Recidiva reacted the same way under intense therapy. Melanomas of the retina were more resistant to this therapy, but after enucleation of the bulbus the tumors did not return (3 to 6 years).

Hodgkins Disease

Our Hodgkins cases were treated individually with enzymes, radiation, cytostatica and corticosteroids. Radiation-resistant cases received only enzymes, systemic and intratumoral. The tumors mostly disappeared promptly.

In cases treated already for 10 to 16 years with enzymes, new tumors and other symptoms yielded always again to enzyme therapy. New tumors appeared mostly after interruption of the enzyme therapy and under special stresses (infection, overwork, nicotin or alcohol abuses). Most patients kept up their work without interruptions.

None of the Hodgkins cases under enzyme therapy developed lymphomas or leukemias.

After the manuscript of this book was already completed and sent to the publishers, the authors received a copy of an interim report by Dr. *Hoefer-Janker* and Dr. *Scheef* of the Radiation Hospital Janker in Bonn. The Janker hospital is the most experienced and most famous German cancer institute. The report, which will follow below, was sent to the German Ministry of Health, Department for Medicare, on request.

Interim report on Enzyme Therapy with Wobe-Mugos in Cancer patients. By H. *Hoefer-Janker* and W. *Scheef*, Radiation Hospital Janker, Bonn, Germany, 21 Dec. 71.

"The therapy with proteolytic enzymes (Wobe-Mugos) has been used in our hospital for over 5 years. We used to inject daily doses of 100 mg i.m. resp. doses of 100-200 mg orally or as suppositories rectally. At first this treatment was performed during 3 months with about 40 patients. We did see impressive palliative results, however, the prognoses of the patients were not changed to any larger extent. Therefore the enzyme therapy was again eliminated from the medicamentarium of our clinic. It must be added, however, that this therapy was performed with far advanced final or prefinal patients only for very few weeks or days. Today we know, that the doses used at the time are completely insufficient in such far advanced diseases.

Induced by reports of a colleague known to us as extraordinarily critical, who stated most surprising regressions of tumors after intratumoral applications of an enzyme preparation, our interest was turned at the beginning of 1969 again to the enzyme therapy with Wobe-Mugos. Only after we could produce a success quota of about 50% by intratumoral injections, with extensive liquefication necroses of tumors or their metastases, and in several cases we were able to bring tumors of the size of "babyheads" which were beyond any further therapy by cytostatics and radiation, to a complete remission, only then we became fully convinced, that this enzyme preparation does not represent just an adjuvans in the tumor therapy, but that it has even to be enclosed in the small group of really causal therapeutics.

Under the impression of our observations we were then also

ready to take up, upon the recommendation of other therapists, the use of these enzymes in the therapy of carcinomatous liquid accumulations in cavities and to test the preparate also under this indication. If the instillate contained at least 500 mg of Wobe-Mugos and the instillations were repeated several times, at the most 24 hours apart, the effusion would completely disappear in more than 50% of the cases resp. recurrences were prevented. With cases where this treatment did not lead to full success, mostly a hemodynamically conditioned fluid accumulation existed, with negative findings of tumor cells in the sediment. In the meantime we have improved the method of application to such extent, that in cases with carcinomatous exudation of the pleural cavity the rate of complete success is much higher than stated above. The instillation into the pleural cavity is rather simple and practically free of dramatic side effects.

The reason why this type of therapy, in spite of the surprisingly splendid results, is being used in the peritoneal cavity in a relatively small number and also why in our clinic it is reserved for only a still small group of patients, lies upon partly really dramatic side reactions which can under conditions reach peritoneal shock. We believe, however, to have soon a therapy scheme on hand which will also prevent with considerable certainty these side reactions. It is understandable that we pay now again also increased attention, under the impression of the observations just described, to the systemic application of the enzyme mixture. But the intratumoral and intracavital application had taught us that an objectivable effect on the tumor itself, which, as mentioned before, led in a number of cases to a full remission, could only be reached by the application of high doses. For the clinical use the postulate for us resulted from this fact, to use enzyme doses, which should lie in an amount at least ten times as high as in our previous systemic treatments. Since the enzyme clysma became available and it was proven by blood chemistry that the enzymes by rectal administration are really absorbed to a great extent, we started a more extensive clinical study whose first results are now on hand.

Since April of this year up to now 79 patients with far advanced cancers of different sites were treated over longer periods with 1 to 4 clysmas daily. Since, understandably, a measurable criterion of success for the effect upon the tumor is still missing in most cases, at first clinical facts for judgement of success have to be brought forth. We hope however that we shall be able to state in about one or two years, in further continuation of this study also on hand of a comparison of the medium survival rate and of those who died meanwhile, as well as on hand of the frequency of recurrences and rate of metastases, a statistically significant difference between the conventionally treated patients and the tumor carriers who were additionally treated with enzymes.

As the first and at the moment most apparent parameter of success of the enzyme therapy the required use of corticoid therapy during the radiation is to be used. All patients at our clinic are treated according to a radiation scheme fixed for each type of tumor. Ray-produced handicaps of the general body condition, as every clinician knows and which go hand in hand with nausea, vomiting, anorexia, diarrhea etc., are successfully controlled by a corticoid if they get beyond the subjectively tolerable amount. The use of cortisone is necessary with about 60% of our patients. The examination of our first 79 case histories revealed that in only 3 cases cortisone had to be administered, whereby it may be added, that the prescription of the cortisone was in hand of a colleague, who knew nothing of the existence of this study and who was entirely directed by the clinical picture and the complaints of the patient.

Furthermore it was remarkable during the study of our casuistics—and this difference seems now already to be statistically significant—that additional ordering of pain-suppressing remedies could almost completely be dropped. In not a single case opiates were ordered.

We are gladly ready to give further detail information to anybody requesting them, but from our side a comprehensive statistical analysis is planned, but not before the conclusion of the presently running study, in about two years. For the moment it can merely be stated with some assurance, according to our up-to-date observations, that the therapy with Wobe-Mugos clysmas definitely increases the tolerance to radiation of the patients, that it reduces his pains beyond any doubt, that it prevents formation of the well known paraneoplastic syndromes almost. totally. The dramatic effects of the enzymes by intracavitar and intratumoral application we discussed already before. In this section of therapeutic endeavors the enzyme therapy occupies already now a firm place in our treatment schemes in our hospital."

Chief physician of the hospital
Dr. W. Scheef

From the Spanish Cancer Institute we also received this letter after this manuscript was already finished.

Spanish National Cancer Institute
of the Ministerium of Health
Gynecological Department
Dr. Lopez de la Osa

Madrid, October 11, 1971

"Gentlemen!
We gladly inform you that our comprehensive clinical investigation with the enzyme preparation Wobe-Mugos and the in megadoses given A-mulsin-high-concentrate have resulted in most surprising positive successes in the therapy of malignant tumors in the gynecological department. We regard it therefore of the greatest importance to be able to further continue these experiments in still greater extent, in order to assure these results with a larger number of patients.

I remain, with friendly greeting,
Dr. Leon Lopez de La Osa"

On December 3rd, 1971, the Austrian Medical Association held their Annual Congress in Vienna with the main theme: "Treatment of the Bronchogenic Carcinoma." Several speakers reported about their surprising positive results in using proteolytic enzymes and megadoses of vitamin A emulsion in bronchogenic carcinoma. Prim. *Titscher,* Chief of the Thorax Department of the City Hospitals of Vienna-Lainz pointed out:
"The resistance to therapy of malignant pleura effusions—caused by bronchuscarcinoma or by metastases—have induced us on the lung department of our hospital, after advice of Prof.

Wrba, to apply Wobe-Mugos intrapleurally. We have at present on hand 8 malignant effusions by bronchuscarcinomas and 38 metastatic pleura carcinoses mainly after mamma carcinomas.

I want to restrict myself here to the 8 cases of bronchuscarcinomas of whom we were able to dry up five within surprisingly short time (average 4 weeks), 2 patients died shortly after start of therapy due to the basic disease, one effusion remained without influence. Among the 5 "dried up" cases were 2 adenocarcinomas. The metastatic effusions responded equally well. At present investigations about the chemically explainable changes in the exudate by the enzyme therapy are going on, as well as histological examinations of the fibrin deposits.

After this Prof. *Heinrich Wrba*, head of the Austrian Cancer Research Institute at the University of Vienna, concluded the congress by stating: "The enzyme treatment of cancer is not only valuable in treatment of carcinoses and exudation of the pleural cavity as we have heard a little while ago, it is also more than a valuable adjuvant in cancer therapy; our present knowledge allows us to include this therapy into the small list of highly effective causal anticancer compounds. It will certainly play an important role in cancer treatment of the coming years. Everyone who is really sincere in his efforts to improve the thus far discouraging results of cancer therapy, has to intensively engage himself and learn about this most interesting and promising new field of therapy."

Directions for the treatment of malignant tumors with proteolytic enzymes:

Our experiences in the therapy with proteolytic enzymes of malignant tumors are limited to the enzyme mixture Wobe-Mugos*, which was developed by us and found optimal especially for this purpose. To what extent similar therapeutic re-

---

*Wobe Mugos: sublingual tablets (25 mg), entericcoated tablets (25 mg), ampoules (100 mg) for i.m. use, suppositories (100 mg), clysmas (1000 mg) and ointments (400 mg in 20 g) manufactured by Mucos GmbH, Gruenwald, Germany.

sults can be obtained with other enzyme preparations available in the USA, and in which dosage, we do not know. Investigations in this respect appear us urgently desirable. It certainly would be best, if an important American pharmaceutical firm would very soon bring into our trade supply the enzyme preparate Wobe-Mugos which has been proven effective in extensive use and eminently successful abroad on many thousands of advanced cancer cases, after our FDA will admit its use in the USA.

General directions: By far the most efficient use is the local or topical application, because thereby a far greater enzyme concentration can act upon the malignant tissue than by systemic use. Proteolytic enzymes are non-toxic to healthy tissues, even in highest concentration. The local application can be performed as intratumoral or peritumoral injections, as intracavitory instillations into the pleura, peritoneum, urinary bladder, or per clysma with colon- or rectum-carcinomas, as vaginal suppositories and as ointments on skin tumors. In case of systemic enzyme applications a large part of the given enzyme is inactivated by the inhibitors of the blood and the tissues, apart from being diluted by the large volume of blood, so that a therapeutic success is only taking place by very high doses.

The local enzyme applications in very high doses aim at the prompt destruction (lysis) of one or more tumors; often one to five applications of 1000-5000 mg suffice to accomplish a total tumor lysis. In case the lysis products cannot discharge toward outside, they must be drained off by punctures.

The systemic use aims, on the one hand, also at the tumor destruction though only by giving very high doses over longer periods, on the other hand at the prophylaxis against metastases and recidiva, furthermore at detoxification of products of tumor lysis, reduction of side effects by radiation therapy and at increase of the effects of other types of therapy, like radiation, cytostatics or vitamin A. At the same time a considerable reduction of the tendency to thromboses and a general roborant effect is accomplished. In cases of prefinal or final stadium of tumor patients one succeeds by high dosage of systemic enzyme use frequently a distinct improvement of the general condition, the requirement of analgesics and opiates is markedly reduced and the patients die mostly without any painful agony.

Doses suggestions for systemic application of Wobe-Mugos enzymes:

Daily 2-4 Wobe-Mugos clysmas, divided in 2 doses 12 hours apart (1000 mg enzyme mixture per clysma dissolved in about 10 ml water) during the first 6-8 weeks resp. during the therapy with emulsified vitamin A, in case this lasts over a longer period. Thereafter 2 times daily 2-3 Wobe-Mugos dragees, enteric-coated or 2 Wobe-Mugos suppositories daily, and monthly for 5 days 1 Wobe-Mugos clysma. This therapy and dosage is also continued after finishing other therapeutic measures, because experience showed that patients who have lived under this enzyme therapy and dosage free of complaints, but still with their tumors, experience after stopping the enzyme therapy an exacerbation or reappearance of their tumor disease within 4-6 months, which then did not respond anymore to any therapeutic measures and led to a fast exitus. If possible, the same doses should be administered in post-surgical and post-radiated patients for the purpose of avoiding the formation of metastases and recurrence. This therapy should be continued for at least 2 years; we feel, however, it is wise to continue for a longer period, maybe for lifetime.

Directions and doses suggestion for intratumoral Wobe-Mugos injections:

Depending on the size of tumor, 1-10 ampules Wobe-Mugos are dissolved in about 2-6 ml physiological salt solution, adding 2 cc of lidocain 1% (which is packed with the Wobe-Mugos ampules) and inject. The injection may, if required, be performed under short-time narcosis. It is wise to use wide and strong needles (as used for puncture of the pleura), because quite frequently the tumors are hard and very solid and it is difficult to push the needle into the hard tumor. It may be necessary to repeat this procedure 3-5 times every second or 3rd day. As soon as the tumor softens it should be tried to puncture and drain the lysed tumor before a new injection is performed. Intratumoral injections have been applied in larger tumors of all abdominal organs and locations, especially of the ovaries, corpus uteri, adrenals and metastases of the liver, and also tumors of the lung, if with certainty the tumor could be reached with the needle. In smaller tumors it may be necessary to do this under X-ray control. Also tumors of other locations, like in the ear, nose and throat area and skin are easily reached with the needle. These intratumoral injections are almost always well tolerated, only in melanomatous skin metastases severe circulatory affections can lead to an acute collapse. It is advisable to have ready injections of pervitin or similar analeptica and to give them slowly i.v., if necessary.

Directions and doses for Wobe-Mugos instillations intrapleurally and intraperitoneally.

After draining the ascites or pleura exudate down to 0.1 to 1 l, 8-10 ampules Wobe-Mugos (800-1000 mg) dissolved in 20 ml physiological salt solution plus 4-6 ml of 1% lidocain are instilled. Possible appearing severe pains can be controlled by analgetica (e.g. novalgin i.v.). After a few instillations which should be repeated every day, the exudates are rendered free of cells and dry up completely. With the instillation of Wobe-Mugos into the peritoneal cavity occasionally a peritoneal shock may ensue which can be rapidly caught with circulation remedies and 1 ampule polamidon and 5 mg novalgin slowly given i.v., or similar analgesics.

The local application of Wobe-Mugos is equally efficient with all malignant tumors, independent of histological structure or location. Sometimes the tumors seem to get larger; in such cases necrotic colliquates can be mostly drained off. In spite of the collapse and shock conditions, which sometimes come about, this method is reliably successful and desirable.

Side reactions: The enteric use of Wobe-Mugos is always free of side reactions. With applications by injections in rare cases allergic reactions resp. light anaphylaxis are caused which can be rapidly controlled by corticosteroids resp. epinephrine. Reactions which are caused by too fast disintegration of the tumor, can be stopped if necessary with protease inhibitors.

Vitamin A in Enzyme Treatment of Cancer

During this period of work, introducing proteolytic enzyme mixtures into various fields of therapy, we were constantly engaged in testing new compounds to improve the effectiveness of our enzymes.

In cooperation with *Hoefer-Janker* and *Scheef* (Radiation Clinic Janker, Bonn) we found megadosis of emulsified vitamin A to be of great value in this respect (26-28).

Megadoses of vitamin A have different pharmacodynamic effects, one of which is a marked liberation of lysosomal enzymes. Systemic enzyme therapy is possible on three different ways:

1. The enzymes, like Wobe-Mugos, become directly effective when absorbed by parenteral or enteral administration. The enzymes penetrate and develop their influence on the substrate directly.

2. The administration of enzymatic mediators, like streptokinase or urokinase, activates enzymogens (inactive enzymes), e.g. plasminogen is converted to plasmin. Also application of trypsin or enzyme mixtures, like Wobe-Mugos, partially activate enzymogens in the blood.

3. By use of various substances, e.g. polyanions, vitamin A and others; the intracellular enzymes located in the lysosomes become activated and are released into the body fluids. Megadoses of emulgated vitamin A proved to have a very high effect in liberating lysosomal enzymes which then cause therapeutic effects.

Among other biochemical compounds, vitamin A became very important lately by investigations of different research centers.

The free vitamin A in physiological amounts stabilizes the membranes of mitochondria and erythrocytes and regulates the permeability in the tissue cells (17, 18, 22, 46). The functioning of the optical perception as well as the synthesis of mucopolysaccharides in the epithelia and skin depend on its presence (33, 34).

Dingle made a significant contribution by showing that vitamin A, given in high doses, destroys the membrane of lysosomes (17, 18). *DeDuve* proposed the name "lysosomes" to a group of cytoplasmatic particles containing a number of acidic hydrolases (19, 21, 23, 29, 36, 39). Thus, lysosomes are cytoplasmatic particles containing numerous hydrolases with various substrate specificities. They are enlcosed by a phospho-protein-lipid-membrane. The following table reviews the enzymes in lysosomes so far isolated, all of which have an activity optimum in the acidic pH-range (14).

Enzyme	Substrate	Author
ribonuclease C.E.2.7.7.17/2.7.7.16		DeDuve
desoxyribonuclease C.E. 3.1.4.6	nucleic acids	DeDuve
acid phosphatase C.E.3.1.3.2		Feigen
cathepsin(s) C.E.3.4.4.9		Finkelstaedt
collagenase C.E.3.4.4.19	proteins	Frankland
phosphoprotein- phosphatase C.E.3.1.3.16		DeDuve
alpha- glucosidase C.E.3.2.1.20	glycogen	Lejeune
aryl-sulphatases C.E.3.1.6.1		DeDuve
β-glucuronidase C.E.3.2.1.31	mucopoly-	
β-galactosidase C.E.3.2.1.23	saccharids	DeDuve
β-N-acetylglucosaminidase C.E.3.2.1.29		DeDuve
alpha- mannosidase C.E.3.2.1.24		DeDuve

According to investigations of *DeDuve* and others, lysosomes differ from other cell organellas by:
1) size of 0.5 micron
2) high amount of hydrolases

3) absence of an oxydative metabolism
4) enclosure by a membrane which is sensitive to different chemical and physical influences.

A bursting or leakage of the lysosomal membrane is brought about by: anoxemia, hypoxemia, pH-changes below the neutral point, X-ray and other ionizing agents and vitamin A overdosage. Cortisol and its derivatives stabilize the lysosomal membranes or have the opposite effect.

On account of the numerous compounds which lysosomal enzymes can liberate vitamin A has met special interest during the last years, not the least by the experimental and clinical investigations of our own scientific groups.

Vitamin A, if given in high doses, labilizes lysosomal membranes. The intralysosomal enzymes are set free after destruction of these membranes and their activity can be measured in the serum or body fluids (22, 23, 42). Vitamin A and other polyene compounds penetrate the lysosomal membranes (17, 18). By electron microscopy a distortion of the membrane can be determined after vitamin A intoxication and a transition of cytomembranes of the coarse-granular structure into a diffuse one (10).

Dingle showed that lysosomes which were isolated from the liver of animals poisoned with vitamin A are unstable.

Animal experiment investigations showed that vitamin A exerts a protective effect against several malignant processes. High doses of vitamin A have a protective action in animals on whom experimental cancers were to be produced on vagina, cervix, forestomach etc. by cancerogenous compounds (5-13, 35, 37, 45, 47).

The development of experimental epithelial malignancies is delayed under vitamin A therapy. With large vitamin A doses also the rate of formation of the Shope papilloma caused by viruses could be reduced (31).

With hamsters a new growth in the bronchi after repeated intratracheal instillations of benzpyrene is almost completely prevented, if the animals receive vitamin A (15, 16). Histological investigations showed that vitamin A prevents the transformation of the ciliary into sqamous epithelia which has to be regarded as the preliminary stage of carcinoma. The table summarizes the results of the animal experiments.

The conclusion can be drawn that vitamin A can prevent the

Chemically induced experimental tumors and vitamin A

author:	carcinogenic compound:	animal:
Rowe	DMBA	hamster
Chu	BP	hamster
	DMBA	hamster
Saffiotti	BP	hamster
Davies	DMBA	mouse
McMichael	Shope papillome	rabbit
Bollag	DMBA	mouse

BP=benzpyrene
DMBA=dimethylamino-benzanthracene

formation of experimental tumors or lead to a remission of already manifest tumors.

The work group about *Brandes* investigated for years the role of the lysosomes in cellular processes, e.g. in malignant growth (2, 3, 10, 11, 15). The authors proved histochemically that the acidic phosphatase in mamma carcinoma cells (mouse) is increased shortly after giving alkylating compounds.

By additional application of vitamin A in large doses the enzyme activity is further considerably increased. In the treated animals intra- and extra-cellular hydrolases were liberated. According to the opinion of Brandes, the antitumor effect of alkylating compounds is accelerated either by increasing deformation of the tumor cells or by a favored transformation into the "active" form.

From these facts it follows that lysosomal enzymes and vitamin A affect tumor growth similar to the effect of proteolytic enzymes. A combination of these anticancer compounds was therefore logical and led to the introduction of a new cancer therapy.

Hoefer-Janker et al. were the first who succeeded in the sensibilization of human malignant tumors with megadoses of emulsified vitamin A which leads to an enhanced effect of alkylating agents, radiation and enzyme therapy (26-28, 38-42).

Within the framework of cancer therapy which justifies some risks, the authors gave knowingly an overdosage of vitamin A. They accomplished therapeutic and pharmacodynamic effects which were unknown up to now and which were also successful in other areas of indications, like hyperkeratosis and psoriasis (28, 30, 43).

In the clinical practice the treatment with vitamin A met first difficulties because a dosage up to the subtoxic limits was impossible to reach with the oily or aqueous vitamin A preparations available commercially. The high level of the vitamin A in the blood necessary for a therapeutic success could only be reached by the use of a special emulsified vitamin A preparation.

The vitamin A level showed a significant increase within 15 minutes after application of emulsified vitamin A (A-Mulsin). The same increase of the vitamin A level of the blood is reached: with A-Mulsin within 15 minutes, with vitamin A aquasol parenteral after 480 minutes, with oily vitamin A orally given also 480 minutes after administration and with oral application of vitamin A aquasol not at all.

It was possible without difficulties to give the A-Mulsin High Concentrate preparation between 60 and 120 mill. I.U. without major side reactions. The vitamin A in the emulsion does not enter to a great extent the liver via vena portae like that of the usual oily or aqueous form, but it enters the blood and lymph circulation in corpuscular form directly over the lymph vessels from the intestinal villi. Only by direct absorption by the lymphatic way into the blood it is possible to flood the body with the therapeutically necessary high vitamin A quantity.

For the past 4 years the sensibilization of malignant tumors for the ray-, enzyme- and chemo-therapy with mega-doses of emulsified vitamin A has been used on more than 35,000 patients at numerous clinics of Germany and other countries. Although the observation time is relatively short, yet certain types of cancer proved especially vulnerable by this combined therapy. Especially favorable therapeutic results are accomplished in cases of squamous epithelial carcinomas at the following locations:

Esophagus, bronchi, tongue, mouth cavity, throat, larynx, penis, portio, vagina, vulva and skin. Also skin metastases of mamma carcinomas respond well to this treatment. A basalcell carcinoma disappears mostly after taking about 30 mill. I.U. vitamin A. Squamous cell carcinoma of the skin can also in many cases be brought to complete remission with vitamin A alone, without radiation or surgery.

Hoefer-Janker et al. demonstrated in a large number of cases that in carcinoma of the bronchi, esophagus and vulva the doses of radiation can be reduced to 50% of the normal range if vitamin A is used. The therapeutic success is still more reliable than if radiation in high doses is used only. We can definitely claim today that vitamin A in megadoses increases the effectiveness of radiation always to a large extent. Also the results of alkylating agents are enhanced by emulsified vitamin A. The combined therapy of vitamin A with proteolytic enzymes like Wobe-Mugos definitely acts synergistically. Our main experience with the use of vitamin A is in squamous cell carcinoma of different locations. Here the therapeutic effects are definitely proven. Also in adenocarcinomas of different locations we have used vitamin A with success, however our experience is still not large enough for final statistical claims. In acute leukemias and sarcomas the effects of vitamin A seem to be poor (26, 27).

In May 1971 the Janker hospital published the results obtained in 37 cases treated with a combination of radiation together with vitamin A of invasive carcinomas of the vulva. In this group of patients treated with the combined therapy the amount of radiation required was reduced by about 1/3 in comparison with those cases receiving radiation only. Almost identical results were observed in a group of esophagus carcinomas where radiation was markedly reduced, compared with patients on radiation therapy alone. The combination treatment produced mostly better therapeutical results than radiation alone.

Since a few years the application of emulsified vitamin A in numerous clinics was investigated.

Herbst of the Urological Dept. (Univ. of Graz) used megadoses of vitamin A in urinary bladder carcinoma. He observed steady tumor regressions in 35% of his patients. However, more important was the result that postsurgical patients showed much less metastases if treated with vitamin A. In patients with bladder papillomatosis he was successful in avoiding recurrence and formation of malignant growth after electrocoagulation in a group of 60 patients over a period of 2 years, by giving vitamin A in megadoses. Also *Adelberger* reports good results in cases with bronchus-carcinomas (11).

At 8 Austrian hospitals under the direction of Prof. *Wrba* (Austrian Cancer Research Institute) emulsified vitamin A was successfully used in inoperable carcinomas of different localizations. In a special study inoperable and generalized bronchial carcinomas were treated with vitamin A high-concentrate and cytostatics or with enzyme therapy (Wobe). Already after an observation time of only 7 months it can be seen that the introduction of the megadoses of vitamin A represents a big progress in the therapy of the bronchus carcinoma. In almost all cases good clinical results could be accomplished with far advanced tumor patients.

There have been several cases of inoperable bronchus carcinoma, verified histologically and by bronchoscopy who died due to thrombosis after intensive therapy with megadosis of emulgated vitamin A. In these patients no chemotherapy with cytostatics or radiation was applied. On obduction it was found that the carcinomas have disappeared completely. Vitamin A has certainly proven its effectiveness in these cases. The cause of death was due to other complications.

All investigators report in agreement that fibrinolysis increases shortly after vitamin A intake. The fibrinolytic activity can be still further increased by simultaneous intake of proteolytic enzymes. The high vitamin A level in the blood also suppresses the keratinization, so that new therapeutic ways are given in the dermatological field (e.g. psoriasis, ichthyosis, morbus Darier etc.).

Ardenne et al. investigated for years the selective damage to cancer cells by a combined attack with local hyperacidity, hyperthermia, vitamin A and other substances which further the liberation of lysosomal enzymes. In numerous investigations the proof was established that damage to the cancer cell not only in vitro, but also in vivo results by combining heat, hyperacidity of the entire tissue and a support of these attacks by vitamin A. These findings were confirmed by *Ollendieck, Berg* et al.

The exact mechanism of action of vitamin A on cancer cells has not as yet been clarified.

Various possibilities of mechanisms are discussed:
1. Liberation of lysosomal enzymes, as pointed out before.
2. DNA-synthesis and mitosis are stimulated. By an overactivity of this reaction, the cancer cells are unable to live due to exhaustion of substrate and energy.
3. vitamin A intensifies the mesenchymal defense.
4. protein synthesis is changed.

Difference of effects of Cytostatics and vitamin A:

Cytostatics:	*Vitamin A:*
inhibit mitosis	stimulates mitosis
inhibit DNA-synthesis	stimulates DNA-synthesis
are effective in transplanted tumors only	is effective in spontaneous and in chemical induced tumors
act as immunsuppressives	acts as immunostimulans
increase cortisone effects	antagonizes cortisone
are carcinogen	is not carcinogen
cause leucopenia	does not cause leucopenia
are non-physiological compounds	is a physiological compound

By comparing the pharmacological effects of vitamin A to alkylating compounds, the immuno-stimulating effect of vitamin A becomes very probable.

In tumor patients there is almost no reaction to BCG vaccination. After megadoses of vitamin A this reaction becomes very strong. This is definite proof for the stimulation of the cellular antibody production which itself may be of great value in tumor therapy.

Homologous skin transplants in rats are rejected after a certain period. The intake of high doses of vitamin A reduces this time. This again indicates a stimulation of cellular antibody production. In respect to the immune approach in cancer therapy much is still unknown and remains to be investigated. Vitamin A in emulsified form is a very useful compound for this investigation. The experimental findings were confirmed by clinical tests (*Hoefer-Janker* etc.).

It is very important to use a system of gradual increase of doses because the organism develops more tolerance by this gradual increase. The therapy with megadosis of vitamin A is not free from side effects. Practically all patients show a complete desquamation of the upper layers of the stratum corneum of the epidermis. During this period some patients are burdened with some itching. Due to increase of liquor formation some patients develop severe headache which is controlled by use of diuretics or oral pain killers.

In almost all patients transaminases are increased at beginning of the treatment but normalize during therapy between 2-8 weeks. In rare cases we found development of adynamia which may develop the symptomatology of myasthenia gravis. In these cases we discontinue treatment until symptoms subside. In rare cases we found the breaking off of hair or nails. All these side effects are only temporary, fully reversible and tolerable. No lasting side effects became known in treatments of more than 35,000 patients. This is also true with non-cancer patients suffering from severe psoriasis, ichthiosis, morbus Darier etc. In all these diseases megadoses of emulsified vitamin A are also of great therapeutic value. Cancer patients tolerate more vitamin A than non-cancer patients. Vitamin A therapy should not be used during pregnancy, glaucoma and severe liver diseases. Corticosteroids act antagonistically to vitamin A and are therefore contraindicated. Also consumption of alcohol in any form must be avoided.

Megadoses of vitamin A cause a stimulation of the immune response, especially cellular antibody formation. Latent infections like those of teeth, tonsils, sinuses or internal organs, even latent rheumatoid and psoriatric arthritis may flare up, but they subside after 2 or more days and remain apparently cured. We have found that proteolytic enzymes and megadoses of emulgated vitamin A are a very important factor in combined cancer treatment with at least some types of tumors. The side effects are tolerable and fully reversible.

With rats which had received different quantities of vitamin A in emulsified form, *Bayerle* determined the status of lipids, some enzymes characteristic for liver functions as well as electrophoretically the albumen bodies in the serum. He sums up the results of his investigation as follows: After giving vitamin A in toxic doses an outspoken hyperlipemia is formed which still persists several more weeks after discontinuation of the vitamin.

Parallel to it goes an increase of the beta-lipoproteids. The hyperlipemia may reach values up to 3 to 5 times the normal and can be found still a month after discontinuation of vitamin A. A significant hyperphosphatidemia takes place while the free fatty acids are lowered. After chronically taking of vitamin A the GOT values are slightly increased although in these cases one cannot talk of a liver damage.

No changes were found in the triglyceride lipase, leucin-aminopeptidase and fibrinogen. Also the status of coagulation, as measured by the time of recalcification and time of native-blood coagulation remained unchanged. The alpha- and gamma-components of the serum proteins can be higher, but regular relations between taking vitamin A and level of serum protein could not be shown on the type of animals used. It has to be emphasized that all described changes are fully reversible and completely disappear without damage to the animals.

Directions for the Treatment of Malignant Tumors with Megadoses of Emulsified Vitamin A.

Our experience with megadoses of vitamin A in malignant tumors is restricted to emulsified vitamin A palmitate (Amulsin-high concentrate)[*], for reasons already discussed in the previous chapter; it was not possible to use other forms of vitamin A.

Indications: Especially favorable results are accomplished with squamous-cell carcinomas (also inoperable) at the following locations: Mouth cavity, Tongue, Throat, Larynx, Esophagus, Bronchus, Nose, Ears, Sinus cavities, furthermore Penis, Portio, Vagina, Vulva, urinary Bladder (also recurring Papillomas of Bladder), Anus and Skin. Also skin metastases of mamma carcinomas, specially carcinoma crysipeloides respond remarkably to this therapy. Smaller basiliomas disappear frequently after application of 30 mill. I.U. vitamin A, even without any other therapy. The same is true for pre-canceroses of the mucous membrane (leucoplakias). With malignant melanomas the therapeutic success is remarkable in many cases though also failures are reported by some clinicians. The mycosis fungoides represents another very successfully tried indication. Adenocarcinomas also respond positively to this therapy, but up to now less comprehensive experiences are available.

---

[*] Mucos Ges m.b.H. 8022 Grünwad, Germany.

Whether also sarcomas are affected by vitamin A is not known yet. Vitamin A megadoses are also very successful in various dermatological diseases, especially in psoriasis, ichthyosis (also congenita), neurodermatitis, morbus Darier, pityriasis, lichen, pemphigenoid, sclerodermia and acne, as mentioned before.

Dosage: Because with this therapy very high doses of vitamin A are required, side reactions cannot always be completely excluded, but they can be reduced to a minimum, if the dose is slowly increased.

For ambulant patients (outpatients) the following dosage is advised:

First day 20 drops = 1 cc = 300,000 I.U. Amulsin-high concentrate
Second day 24 drops = 360,000 I.U. Amulsin-high concentrate
Third day 28 drops = 420,000 I.U. Amulsin-high concentrate
and every following day the dosis is increased 4 drops. On the 21st day therefore the dose of 100 drops (= 1.5 mill. I.U. vitamin A per day) is reached. If at this date the skin peeling or desquamation reaction, as usually expected, has begun, it is advisable to continue this dose daily till the contents of the container has been used up. The continuation of the therapy (second container A-Mulsin-high concentrate) with the same doses (100 drops daily) depends upon therapeutic results and is left to the individual decision. Experience shows that the peeling reaction is in some cases retarded; then it is advised to continue increasing the dose daily with 4 drops (2nd and 3rd container) up to the asbolutely maximal dose of daily 200 drops (10 ml = 3 mill I.U. vitamin A) which, if steadily continued, is reached on the 46th day and is then continued till the third container is used up. The daily dose may be taken at once or in divided amounts during the days, the patient should individually test which mode agrees most with him.

For hospitalized patients the following dosage is advised:
First day 20 drops = 1 ml = 300,000 I.U. Amulsin-high concentrate
Second day 30 drops = 1.5 ml = 450,000 I.U. Amulsin-high concentrate
daily increase 10 drops until absolute daily maximal doses of

200 drops = 10 ml = 3 mill. I.U. vitamin A is reached on the 19th day. At the same time the first container AMH is used up. The therapy is continued without exceptions with daily 200 drops until 2 containers (60 mill. I.U.) or 3 containers (90 mill. I.U.) are used up. In the most difficult cases the contents of up to 5 containers AMH may be given divided over a proper period of time under extremely careful monitoring. In order to avoid recurrences, a daily sustaining dose of at least 20 drops (= 300,000 I.U. vitamin A) is desirable, even if a complete remission has been accomplished. This sustaining dose should be continued for at least three months, in many patients it was continued for 1-3 years. *Absolute prohibition of alcohol is urgently required.*

Course of therapy: Typical for the therapy with AMH is the moulting of the skin which starts individually different, usually between the seventh and twenty-first day. This shedding of the upper skin layer in large pieces (like after a sunburn) is frequently connected with rhagades of the lips and nostril lining, which sometimes also lead to slight bleeding.

Contra-indications: Existing or possible pregnancy, glaucoma, severe hypertonia, severe diabetes, liver- or kidney diseases. Young people below 18 years of age should be treated with AMH only under rigid clinical control because vitamin A megadoses could affect cartilage or bone formation.

Vitamin A in megadoses is also an antagonist of the corticosteroids. The simultaneous addition of corticosteroids therefore interferes with the desired success. The same seems to be true also for phenylbutazon and its derivatives as well as antiphlogistica with corticoid-resembling effect. Medication given to increase blood pressure is best interrupted.

Antiovulatory medication should be given to women who may get pregnant during this therapy. Whether the antagonism between vitamin A and estrogen neutralizes the prevention of ovulation could not yet be clarified with certainty, but there are no reports to the contrary either. Diabetics should continue their therapy as before but should be particularly strictly supervised.

According to experiences in more than 35,000 patients, sometimes side reactions appear during the therapy with AMH. These should not cause a premature discontinuation of the therapy, because they are controllable and reversible by proper measures as advised below:

Symptoms	Etiology	Suggested Therapy
1. Headache	slight increase of liquor pressure	analgesica, diuretica, reduction of AMH doses
2. Nausea, vomiting	increased liquor pressure	diuretica reduction of AMH doses
3. Cheilitis (splitting of lips)	part of the peeling reaction	cortisone ointment
4. Cold sensation, shivers	hypo thyroid syndrome	0.5-1 gr thyroid tablet/day
5. Nervous restlessness	hypo thyroid syndrome	0.5-1 gr thyroid tablet/day
6. Breaking or loss of hair and nails	thyroid irritation	no therapy, reversible before 6-9 months
7. Hypotonia	latent arthritis	table salt, raise legs
8. Joint pains	cortisone antagonism	analgesica, interrupt therapy
9. Photophobia	increased sensitivity	sunglasses
10. Urticaria	beginning peeling	warm baths, oils and antihistamins eliminate cause
11. Deafness	activation of latent focus, earwax	reduce dosage of AMH
12. Latent infections activated	latent focus	
13. Hyperergic skin reactions	cortison-antagonism	clears up without therapy

Symptoms	Etiology	Suggested Therapy
14. hemorrhagic diathesis	harmless component of symptoms	no therapy
15. generalized bleeding tendency	doses is too high	interrupt therapy, vit. K
16. severe intolerable side effects	doses too high, idiosyncrasies	stop treatment immediately, 200 mg prednison i.v.
17. myasthenia, adynamia somnolence	hyperplasia of Thymus	stop treatment immediately, prostigmine, mestinone
18. muscle weakness	temporary thymus hyperplasia	Calcium, vit. D

Several pathological laboratory values sometimes appearing:

Elevated BSR	focus activation	no therapy
slightly elevated transaminases	cell moulting	no therapy
slight hemolysis, bilirubin call . . .	cell moulting	no therapy
bromthalein retention elev.	vitamin A overwork of liver	no therapy

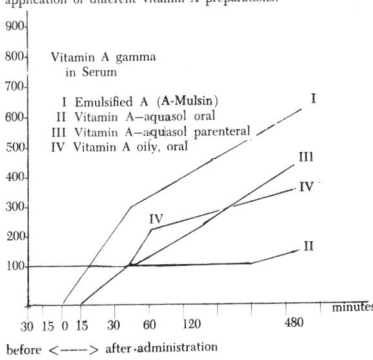

The increase of the vitamin A level in the serum of dogs after application of different vitamin A preparations:

Combination with radiation: With cases suitable for radiation, this therapy should be given simultaneously with the AMH therapy. The effects of radiation are markedly increased, side effects, however, reduced.

Combination with cytostatics: A shock dose of e.g. cytoxan should be given—if indicated—on the last day of AMH therapy (after 60 or 90 mill i.u. vit. A).

Combination with enzyme therapy: During the entire duration of the therapy with vitamin A, proteolytic enzymes (Wobe-Mugos) in doses as high as possible are absolutely desirable. (See directions for enzyme treatment of cancer page 201 etc.)

Literature

1. Adelberger, L., Z.f. Erkrank. d. Atmungsorgane 134, 407, 1971
2. Anton, E., etc., Exp. molec. Pathol. 7, 1561, 1967
3. Anton, E., etc., Cancer 21, 483, 1968
4. Ardenne, M. von, in: Therapeutische und exp. Grundlagen der Krebsmehrschritttherapie, Berlin, 1970
5. Bollag, W., Int. Z. Vitaminforsch. 40, 299, 1970
6. Bollag, W., Schweiz. med. Wschr. 11, 11, 1971
7. Bollag, W., etc., Schweiz. med. Wschr. 101, 11, 1971
8. Bollag, W., etc., Agents a. Actions 1, 172, 1970
9. Bollag, W., Experientia 27, 90, 1971
10. Brandes, D., etc., Lab. Invest. 15, 987, 1966
11. Brandes, D., etc., Cancer Chemother. Rep. 50, 47, 1966
12. Chu, E. W., etc., Cancer Res. 25, 884, 1965
13. Cohen, M. H., etc., Proc. Amer. Ass. Cancer Res. 10, 14, 1969

14. Daems, W. Th., Mouse liver lysosomes and storage, Luctor a. Emergo, Leiden, 1962
15. Davies, R. E., etc., Fed. proc. 24, 2396, 1965
16. Davies, R. E., Cancer Res. 27, 237, 1967
17. Dingle, J. T., Biol. Rev. 40, 422, 1965
18. Dingle, J. T., Lysosomes, Ciba Found. Symp. ed: A. V. S. de Reuck, Boston, 1963
19. DeDuve, Chr., in: Subcellular Particles, ed: T. Hayashi, Ronald, N.Y., 1959
20. DeDuve, Chr., Biochem. J. 73, 610, 1959
21. DeDuve, Chr., in: Lysosomes, ed: de Reuck etc., Boston, 1963
22. Fell, H. B., Vit. a. Hormones 22, 81, 1964
23. Frimmer, M., etc., Int. Z. Vitaminforsch. 38, 454, 1968
24. Frost, G. D., etc., J. Amer. Med. Ass. 207, 1863, 1969
25. Fulton, J. E., etc., Arch. Derm. 98, 396, 1968
26. Hoefer-Janker, H., etc., Arztl. Praxis 23, 538, 542, 1969
27. Hoefer-Janker, H., etc., Kreosarzt 24, 203, 1969
28. Hoefer-Janker, H., Arztl. Praxis 23, 2805, 1971
29. Lucy, J. A., etc., Nature 204, 156, 1964
30. Luger, A., etc., Z. Haut-u. Geschl.-Krkh. 46, 133, 1971
31. McMichael, M., Cancer Res. 25, 947, 1965
32. Nathanson, L., etc., J. Clin. Pharmacol. 9, 359, 1969
33. Niemann, C., etc., Vitamins a. Hormones 12, 69, 1954
34. Pitt, G. A. J., Int. Z. Vitaminforsch. 35, 249, 1965
35. Polliack, A., etc., Cancer Res. 29, 327, 1969
36. Rupec, M., Z. Haut-u. Geschl. Krkh. 45, 67, 1970
37. Saffiotti, U., etc., Cancer 20, 857, 1967
38. Wolf, M., etc., Cancer Congress, Mexico, 30, 9, 1971
39. Ransberger, K., Int. Congress of Chemotherapy, Prag, 1971
40. Hoefer-Janker, H., Vitamin A and Cancer, Österreichische Krebsforschungsgesellschaft, Wien, 21, 1, 1971
41. Wolf, M., etc., Congress, Malta, May 1971
42. Ransberger, K., etc., X. Int. Cancer Congress, Houston, Tex., USA, 27, 5, 1970
43. Kligman, A. M., etc., Arch. Derm. 99, 469, 1969
44. Day, E. D., The Immunochemistry of Cancer, Thomas, Springfield, Ill., 1965
45. Zilber, L. A., Neoplasma 6, 337, 1959
46. Baxter, I. G., Fortschr. Chemie org. Naturstoffe, 9, 41, 1952
47. Dowwing, J. E., etc., Proc. Natl. Acad. Science 46, 587, 1960

IN CONCLUSION

In this monograph the authors tried to refer to the present state of our knowledge about enzyme therapy whereby the particular emphasis was laid on the desire to show the practicing physician the connections between basic facts and their application in his practice. According to the very extensive literature so far available, there seems to be no doubt that enzymes, especially proteases, represent pharmaceuticals of great versatility of effects and of extraordinary importance. In numerous diseased conditions this group of substances proved useful and effective. The pharmacological principle of activity has been clarified to a great extent, also the therapeutic modus operandi is known. The important advantages of the treatment with proteolytic enzymes or physiological body-proper substances have been demonstrated thoroughly. It is to be presumed that enzymatic therapy, especially for cases of malignancy, will extend its application in steadily increasing therapeutic areas and that by the systemic use of enzymes the final limits and possibilities of such treatments will be determined. The same is true for physics of aging.

With every trial of evaluation of therapeutic results it must not be overlooked that every diseased condition does not run like a chemical reaction in preformed channels according to strict fundamental laws, but is governed by many individual factors. This fact determines to a considerable extent any therapy, never mind in which disease. Therapeutic effects do not depend only on the use of the remedy, but just as much on the condition of the patient. This knowledge must prevent us from rejecting in a lump a form of therapy which is relatively new, or holding back its benefits to the patient. It stands to reason that not every disease is curable, but nobody lives forever and the very knowledge that an unavoidable end can be changed into an easily endurable condition, should induce the physician to take advantage of all possible means.

We believe that from this point of view and encouraged by the plainly visible results of enzyme therapy, we may derive the justification to communicate to the physician and the student the necessary fundamental knowledge in the present form.

* * *

The authors are aware of their duty to answer any question regarding these problems as best we can. Suggestions and comments by our readers will be listed in future editions.

ENZYME THERAPY

The following characteristic pictures of cell cultures show the results on normal and on malignant cells under the influence of the enzyme mixture Wobe-Mugos. While healthy tissues keep on growing without interference (Fibroblast cultures Fig. 1 and 2) by the addition of the enzyme mixture, cancer tissues (Fig. 3 to 8) are already dissolved after a short time.

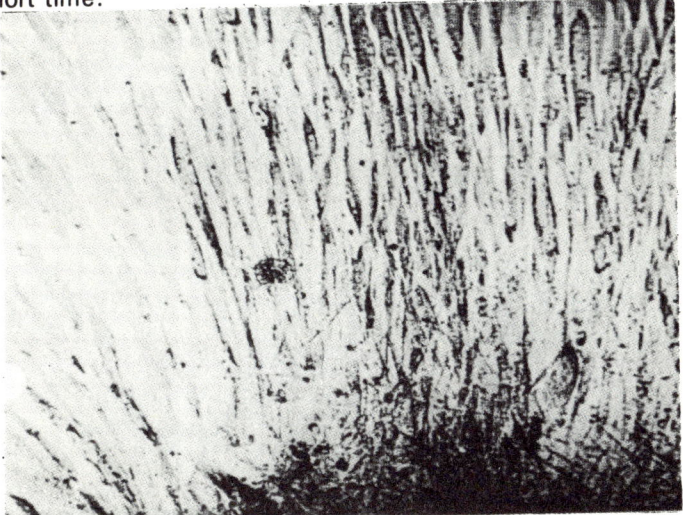

Fig. 1 and 2: Normal fibroblast culture 2 hours and 24 hours after addition of 15 Gamma enzyme mixture. Active growth of all cells.

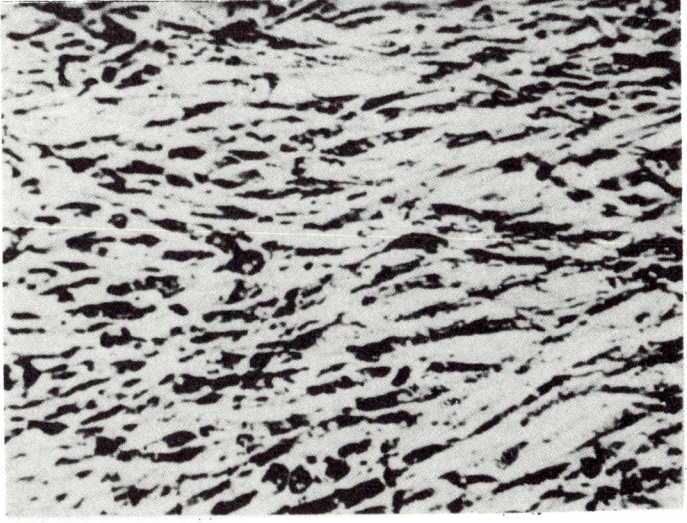

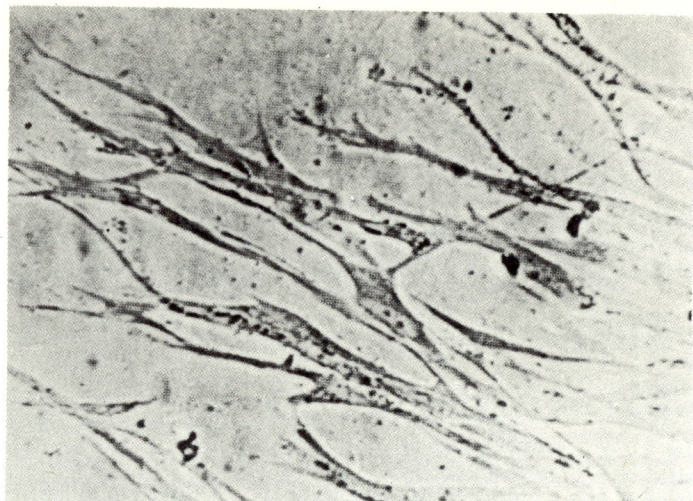

Fig. 3: Cancer tissue (lympho sarcoma) shortly after addition of the enzyme mixture. Vivid growth.

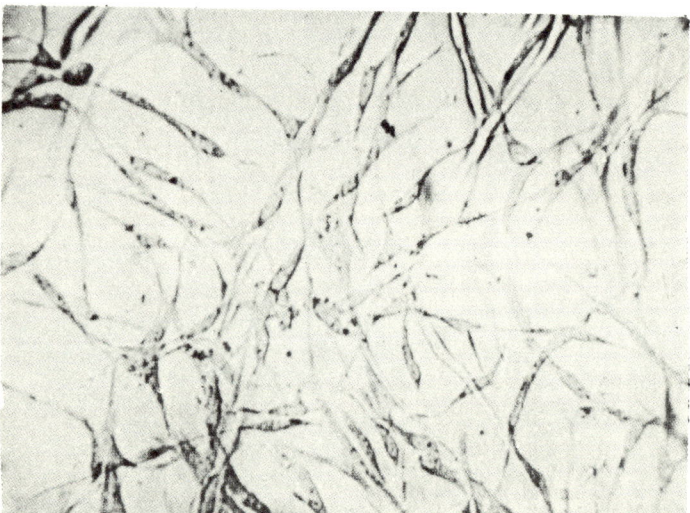

Fig. 4: The same cellculture like Fig. 3, two hours after addition of 15 gamma enzyme mixture.

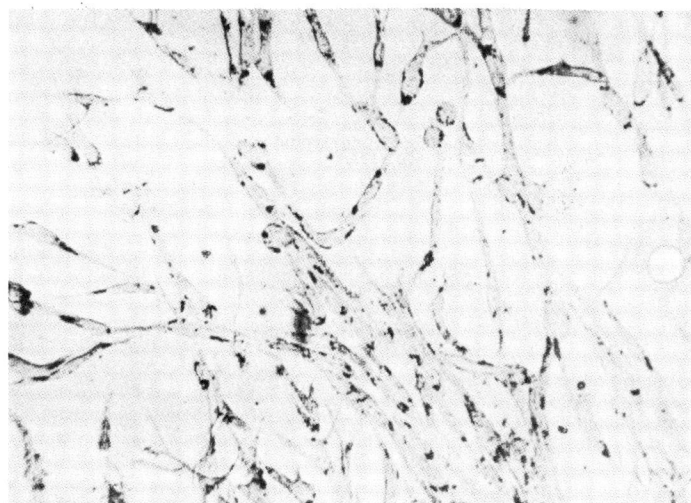

Fig. 5: Six hours after addition of 15 gamma enzyme mixture. Several enucleation become visible (round forms).

Fig. 6: 12 hours after addition of the enzyme mixture. No signs of cell life anymore, extensive disintegration of all cells.

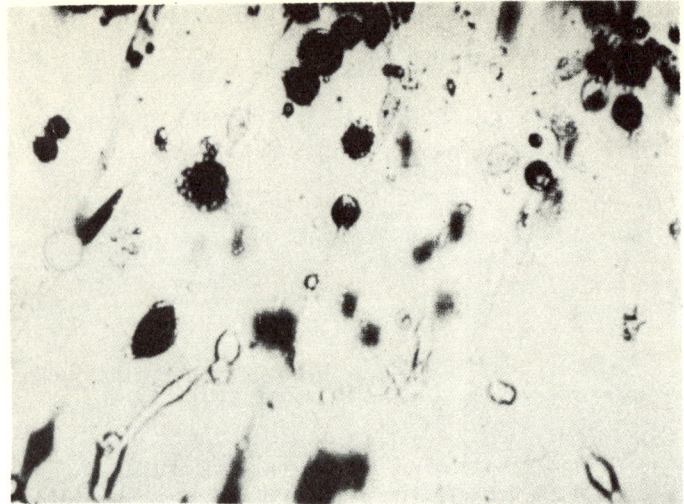

Fig. 7: (high magnification) 12 hours after addition of 15 gamma enzyme mixture: Enucleation and destruction of almost all cancer cells. Only nuclei are visible, the cytoplasm almost completely dissolved.

Fig. 8: Same slide, 24 hours later. Complete cytolysis.

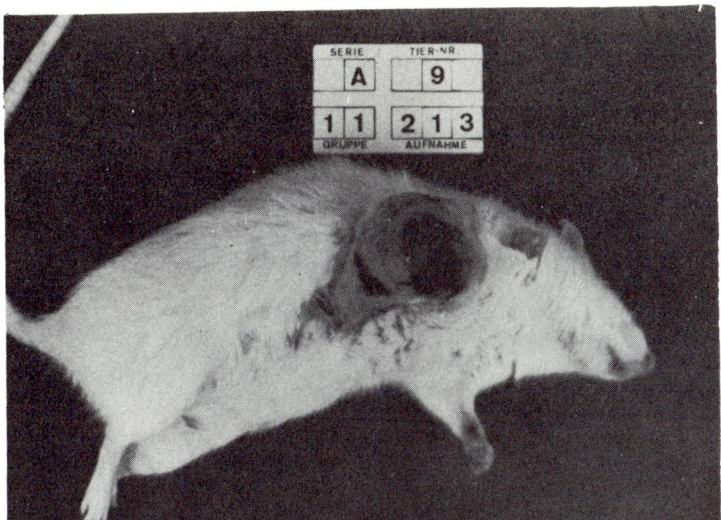

Picture 1: Rat, 30 minutes after getting 100 mg. proteolytic enzyme intratumorally: Short time after the injection tumor shows spreading hemorrhagias.

Picture 2: Same rat 24 hours after above injection: The hemorrhage increased, the tumor surface begins already to become necrotic.

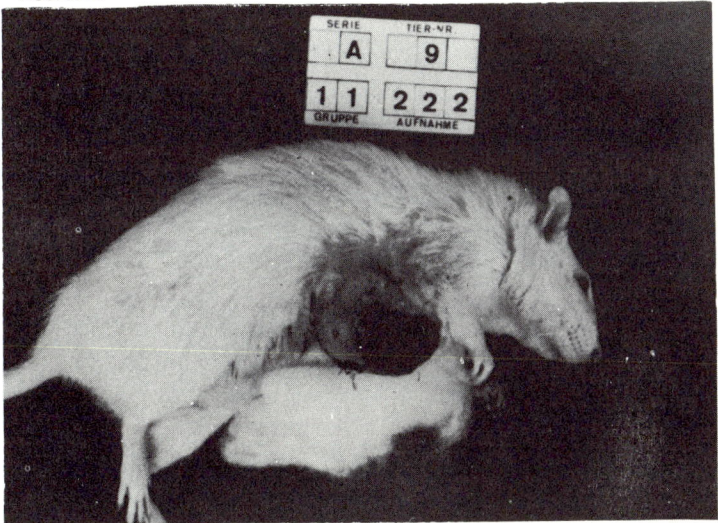

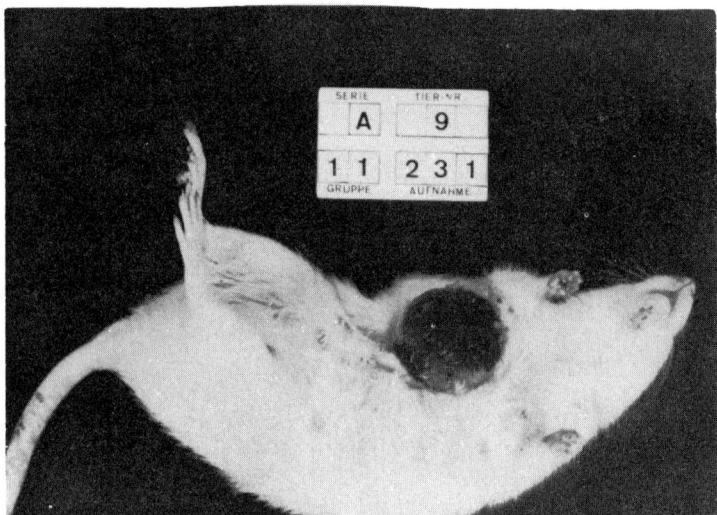

Picture 3: Same rat after 48 hours: After a second injection of 100 mg. proteolytic anzymes intratumorally: The active tumor is fully hemorragic, the necroses are spreading.

Picture 4: Same rat after 96 hours: After a third injection of 100 mg. enzymes intratumorally: The necrotization processes continue, a part of the tumor has been cast off, the process continues partly as dry necrosis.

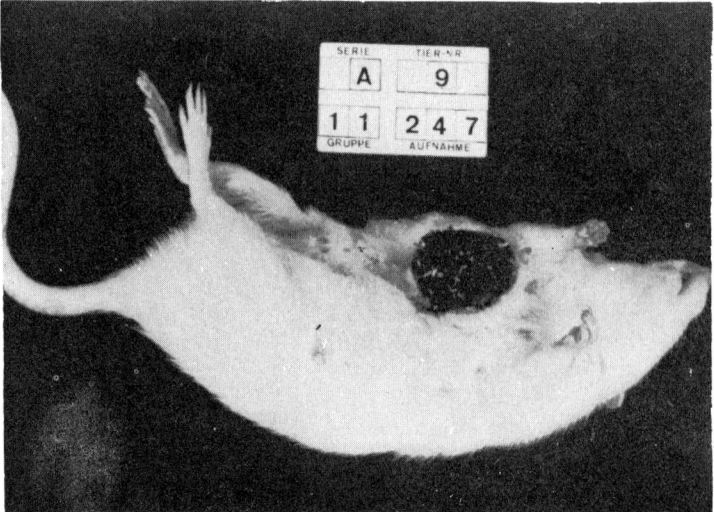

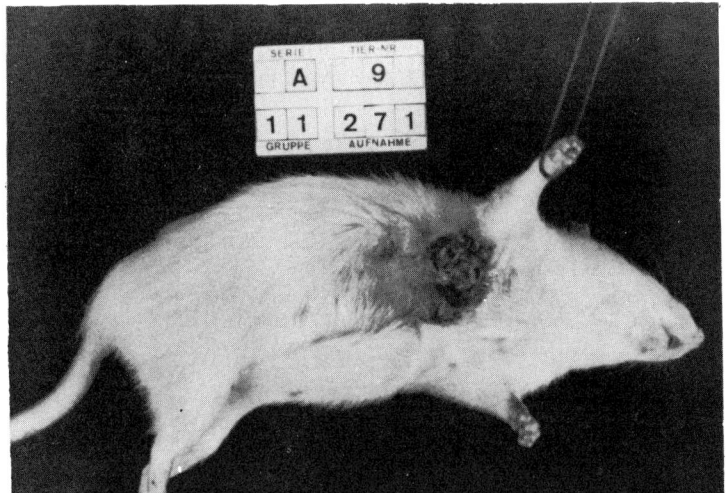

Picture 5: Same rat after 240 hours, after 4 injections of 100 mg. proteolytic enzymes each and twice application of enzyme salve. The hemorrhagic—necrotic tumor has been completely cast off. The picture shows a slightly inflammatory wound area.

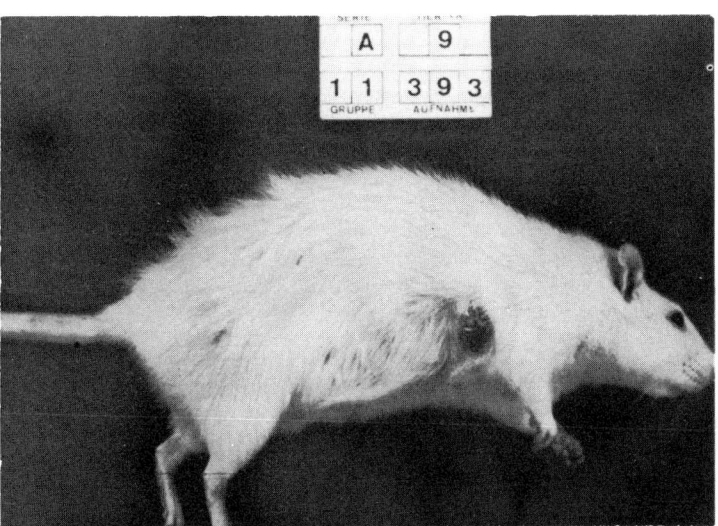

Picture 6: Same rat after 672 hours (28 days) after totally 4 injections of 100 mg. proteolytic enzymes, 4 application of enzyme salve and 3 times locally applied anti biotica: Almost complete regeneration of the wound area; meanwhile the animal has fully recovered, the wound area fully overgrown with hair and no scars are palpable anymore.

INDEX

A

AHF = (Antihemophilic factor) 49
AMH, 216, 217, 219
ATEE = (N-Acetyl-L-Tyrosine-ethylester), 29
A-Mulsin, 209, 214, 215
Abderhalden, 24, 46
Abercrombie, 152, 154, 164
Accelerin, 49
Acid phosphatase, 206, 208
Activator, 19, 41, 53, 63, 65, 79, 80, 128, 139
Adelberger, 98, 211, 219
Adhesion, 155
Aging, 104
Agostino, 151, 156, 157, 158, 164, 165
Agouti, 148
Aliksir, 13
Alpha-glucosidase, 206
Alpha-mannosidase, 206
Ambrose, A. M., 25
Ambrose, E. J., 133
Ambrus, 79, 151, 164
Amine, biogene, etc., 41, 43
Amris, 58
Amylases, 21, 24, 145
Amulsin, 215
Amylopsin, 60, 135
Andreenko, 65, 68
Andrews, 134
a-1-Antichymotrypsin, 56
Anticoagulant, 74, 75, 86, 87, 88, 157, 159, 160, 161, 162, 163
Anticoagulant therapy, 74, 162
Antihemophile Globulin A, 49
Antihemophile Globulin B, 49
Antihemophilic factor = (AHF), 49
Anti-inhibitors, 19, 144
Antiplasmin, 56, 71
a-2-Antiplasmin, 56
Antiprotease, 26
Antistreptokinase, 70, 73, 76
Antithrombin, 31 56
a-1-Antitrypsin, 56
Anton, 219
Apoenzyme, 16, 19
Ardenne, 211, 219
Arginase, 147
Aryl-sulphatases, 206
Aschoff, 61
Ashworth, 150, 164
Asparaginase, 147, 148, 149
Aspergillus oryzae, 23, 141
Assay, 18
Astrup, 39, 40, 46, 54, 56, 58, 59, 61, 63, 68, 80, 82, 106, 112, 117, 141
Atheroma, 111
Atherosclerosis, 54
Auto-pro-thrombin II, 49
Autoimmunity, 115
Autothrombin I, 49
Avakian, 30, 32

B

BAEE = (Benzoyl-arginin-aethylester), 28, 29
BCG vaccination, 212
BAEE = (Benzoyl-arginine-ethylester), 28, 29
Babson, 28
Back, 66, 68
Bacteriophage, 119, 120, 121, 122

Bacteriotropin, 37
Bainbridge, 136, 146
Bale, 156, 165
Bamann, 102, 103
Bare, 28, 32
Barnett, 32
Barri, 131, 133
Barrows, 111, 117
Barth, 45, 46, 175
Bartsch, 182
Basal cell carcinoma
 (Basalioma), 196, 210
Baserga, 151, 164
Baumer, 76
Baxter, 220
Bayerle, 129, 133, 213
Beard, 135, 136, 146, 167, 170
Beiler, 32
Bein, 53
Belitsev, 32
Benda, 137
Benitez, 126, 128, 134
Benzoyl-arginine-ethylester =
 (BAEE) 28, 29
Benzoyl-phenylalanine-
 naphthyl-ester = (BPNE), 29
Berg, 211
Bergmeyer, 25
Bernard, 14
Berzelius, 14
B-galactosidase, 206
B-glucuronidase, 206
Bhuyan, 159, 166
Bieling, 126, 133
Bier, 22
Biggs, 62, 68
Biochemistry of Enzymes, 21
Biological Research Institute,
 128, 132, 139
Birren, 117
Bjerrehuus, 80, 82
Bladder carcinoma, 177, 179, 187,
 192, 202, 210
Bladder papilloma, 192

Blood Coagulant, etc., 162, 163
Blumenthal, 136, 146
Boeryd, 161, 165
Boissonnas, 32
Bollag, 116, 117, 219
Bonifas, 134
Bostrom, 75, 76
Brandes, 219
Bradykinine, 31, 41, 42, 43
Bromelin, 21, 27, 31, 44, 141, 187
Bronchial carcinoma, 172, 182,
 184, 194, 195, 200, 201, 211
Broome, 148
Bross, 73, 76
Brown, 126, 134
Brown-Pearce Carcinoma, 158
Buchner, 14, 61, 67
Buluk, 80, 82
Burnett, 115, 118
Bussy, 14
Bynoe, 134

C

Ca-factor, 49
Caldwell, 81, 83
Calcium chloride, 49, 155
Campbell, 136, 146
Carboanhydrase, 15
Carcinogen, 152
Carica papaya, 23
Carlson, 114
Carzodelan, 139, 143, 144
Cathepsin, 23, 206
Celander, 80, 82
Cepelak, 62, 67
Cervix carcinoma, 183, 185
Chambers, 32
Charité, 139
Chemosis, 45
Chemotaxis, 37, 42
Cholesterol, 54, 138
Christenson, 47, 53, 58
Christiani, 138, 146

231

Christofani, 127, 133
Christmas-factor, 49
Chu, 219
Chymotrypsin, 14, 20, 21, 26, 28, 29, 30, 32, 44, 53, 55, 60, 64, 67, 126, 139, 141, 145, 187
Chymotrypsinogen, 19, 21
Clark, 63, 67
Cleaves, 136, 146
Cleeland, 126, 133
Cliffton, 46, 78, 79, 155, 156, 160, 161, 165
Co-enzyme, 16, 19
Co-factor, 49
Cohen, 31, 58, 66, 68, 219
Cohesion, 158
Cole, 164
Collagen, 39, 109, 110, 111, 115, 139
Collagenase, 206
Collum carcinoma, 191
Colon carcinomas, 193, 194
Coman, 127, 133, 153, 154, 164
Convertin, 49
Corpus-Carcinoma, 185, 191
Co-thromboplastin, 49
Cottier, 134
Crosslinks, 111
Curtfield, 136, 146
Curtis, 106
Cytozyme, 49
Cytostatica, 187, 196

D

Daems, 219
Darzynkiwieckz, 32
Dastre, 47
Daum, 175, 190
Davies, 219
Day, 156, 165, 220
Dechtjar, 79
De Duve, 205, 206, 219
Denis, 47, 52
Dentistry, 98

De Sa, 183
Desmosines, 111
Desoxyribonuclease, 187, 206
Desoxycholic acid, 155
Deutsch, 163
Dicoumarol, 159, 160
Didesheim, 32
Dietz, 32
Dingle, 205, 207, 219
Diniz, A. R., 32
Diniz, C. R., 32
Dirr, 102, 103
Dixon, 21, 24
Dormant state, 151
Dorrer, 175, 176
Douglass, 135
Dowwing, 220
Drance, 79
Dresser, 116, 118
Dunkel, 129, 133
Duprey, 136, 146
Dyck, 32

E

Eclipse, 122
Eddy, 84, 99
Edson, 63, 67
Ehrich, W. W., 46
Ehrlich-Carcinoma, 144
Eichenberger, 160
Elastin, 110
Emeterio, 183
Endopeptidase, 21, 55
Engell, 164
Enterokinase, 21, 22
Enzyme-activator, 17, 19
Enzyme Anti inhibitors, 19
Enzyme-Inhibitor, 18, 19
Enzyme Resorption, 26
Epithelial carcinoma, 162
Epitheliomas, 162, 183, 196
Enzymopathy, 13
Erdös, 32
Escherichia-coli, 148
Euglobulin lysistime, 78

Evans, 78, 79
Extrinsic system, 48

F
FDA, 202
FSF, 50
Factors of coagulation, 49-50
Factor III, 62
Fantoni, 90, 91, 95, 99
Fell, 195, 219
Ferlazzo, 134
Fibrin, 38, 39, 40, 43, 44, 49, 50, 51, 52, 53, 54, 55, 56, 58, 59, 64, 65, 66, 69, 75, 80, 97, 112, 115, 127, 128, 156, 163, 180, 201
Fibrin formation, 40, 112, 151
Fibrin-stabilizing factor, 50, 59
Fibrinase, 59
Fibrination, 163
Fibrin net, 42, 153, 154, 156
Fibrinogen, 39, 40, 48, 49, 50, 51, 52, 55, 56, 57, 58, 59, 65, 114, 115, 156
Fibrinokinase factor, 50
Fibrinolysis, 21, 40, 47, 48, 50, 51, 52, 54, 55, 58, 59, 61, 62, 63, 66, 71, 80, 112, 113, 114, 156, 157, 161, 162, 170, 179, 180
Fibrinolytica, 53, 64, 67, 157, 160, 161
Ficin, 141, 187
Finkelstaedt, 206
Fischbacher, 71
Fischer, 74, 150, 152, 164, 165
Fisher, 84, 85, 99, 157, 158, 165
Fleischhacker, 71, 76
Fletcher, 70, 71, 72, 73, 76, 77, 81, 83
Fluroescent dye, 27
Folk, 32
Frankland, 206
Franz, 96, 99
Fred, 66, 68
Freund, 137, 144, 146, 167
Freihofer, 177
Freund-Kaminer, 137
Frey, 80, 81, 82
Friedl, 129, 134
Frimmer, 220
Frost, 220
Fclgrave, 94, 95, 99
Fulton, 220
Fungal protease, 187

G
Gaschler, 139, 143, 146
Gastpar, 150, 154, 163, 164
Gastric carcinomas, 193
Genital carcinoma, 162
Gilfoil, 32
Glock, 129, 134
Goeth, 136, 146
Gorini, 22
Gosemarker, 182
Gottlob, 65, 68
Götz, 127, 130, 134
Gowans, 152, 166
Graebner, 45
Gram, 57
Green, 32
Grimminger, 96, 100
Gross, 66, 68, 72, 74, 76, 163
Grossi, 157, 158, 164
Gsell, 126, 133
Gullick, 32
Gynecology, 95

H
Hadfieldt, 152, 164
Hagemann factor, 50
Hahn, 113
Hald, 136, 146
Hammond, 58
Hansen, 81, 83
Hardy, 84, 99
Hartet, 80, 82
Hashimoto, 81, 83
Haustein, 32
Hayflick, 136
Heide, 32

Heine, 134
Heinessen, 134
Heister, 46
Hendley, 32
Heparin, 65, 79, 88, 113, 116, 157, 158, 163, 187
Herbst, 210
Herpes simplex, 130
Herpes zoster, 129
Herzer, 177
Hess, 72, 74, 75, 76
Hestrin, 28, 32
Hey, 64, 68
Hiemeyer, 65, 68, 71, 73, 74, 76, 158, 161, 165
Hippocrates, 47
Hiramoto, 165
Hirsh, 75
Histaminase, 41
Histamine, 41, 42, 43, 44, 61
Hitchock, 134
Hodgkins Disease, 196, 197
Hoefer-Janker, 116, 118, 174, 175, 197, 205, 208, 210, 212, 220
Hoelzl-Wallach, 142
Hoff, 49
Hoffman-Ostenhof, 25
Holo-enzyme, 16
Homeostasis, 42
Hoof-and-Mouth-disease, 122, 126
Horejshi, 182
Howell, 75
Howland, 160, 166
Hummel, 28, 33
Hungarian factor, 50
Hydrolases, 17, 205, 206, 208
Hypernephroma, 172, 177, 192
Hyaluronidase, 152

i
Immobilized enzymes, 19, 20
Indigestion, 102
Inert protein, 22
Inflammatory substances, 35
Inhibitors, 17, 31, 35, 36, 44, 52, 56, 59, 60, 62, 70, 91, 93, 98, 108, 127, 138, 143, 149, 161, 162, 179, 202, 204
Innerfield, 26, 28, 30, 3, 33, 44, 46, 47, 53, 65, 66, 68, 84, 85, 93, 94, 99, 100
Inter-a-Trypsin-Inhibitor, 56
Interferon, 125
Intrapleural, 201, 204
Intertumoral, 202, 203, 204
Intrinsic system, 48
Isaacs, 125, 134
Isomerases, 17
Isoenzymes, 19
Iwanowski, 122
Iwasaki, 151, 164

J
J-131, 26
Januszko, 80, 82
Jenkins, 84, 99
Jenson-Sarkoma, 147
Johnson, 68, 73, 77, 80, 82
Jonas, 57
Jonasson, 152, 164
Jürgens, 61

K
Kabacoff, 28, 29, 30, 33
Kaderabek, 91, 99
Kahn, 72, 74, 76
Kallikrein, 43
Kallos, 28, 33
Kaltenbach, J. P., 159, 165
Kaminer, 167
Kaminura, 33
Kamiya, 64, 68
Kaplan, 53
Kathepsin, 141
Kaulla, 80, 82
Kay, 27, 33
Keim, 175
Kellner, 33
Kidd, 151
Kinin, 31, 41, 42, 44

Kininase, 44, 52, 53
Klein, 137
Kline, 33
Kligman, 220
Klose, 178
Klücken, 102, 103
Koeppel, 51
Koike, 157, 165
Kojima, 155, 165
Koller, 72, 76
Kopp, 75, 76
Körtge, 72, 76, 77
Kraut, 31, 165
Kretz, 137, 146
Kropp, 80, 82
Kryles, 84, 99
Kühne, 14
Kunitz, 14, 137
Künzer, 65, 68

L

L-asparaginase, 148
LL-factor, 50, 59
LPF-factor, 38
Labile factor, 49
Lactose-dehydrogenase, 16
Ladehoff, 80, 82
Laki-Lóránd-factor (= LL-factor), 50, 59
Lantz, 95
Lasch, 163
Laufmann, 84, 99
Lautz, 95, 100
Lauwers, 33
Lawrence, 157, 158, 165
Lejeune, 206
Lens esculenta, 141
Lester, 57
Leucocytes, 37, 38, 41, 105, 150, 153
Leucocytosis, 37, 38, 43, 73
Leucopenia, 38, 173
Leucotaxins, 36, 37
Leukemia, 130, 186

Leukoplakia, 183, 196, 214
Levin, 152, 164
Lewis, 58
Lewis, Va., 158
Lewis Sarcoma, 158
Lichtmann, 94, 100
Liebig, 14
Ligases, 17
Lindemann, 134
Lindner, 36, 46
Lionell, 159
Lipase, 14, 16, 21, 23, 145, 187
Lisicky, 91, 99, 180, 181
Lisnell, 166
Little, 136, 146
Liver-catalase, 141
Lohmüller, 177
Lombardo, 134
Long, 152, 164
Lopez da Osa, 184
Lucy, 220
Ludwig, 65, 68, 162, 163, 165, 166
Luger, 220
Lung-Carcinoma, 186
Lüscher, 49
Lustig, 137
Lyase, 17
Lymphogranulomatosis, 130, 151
Lysis, 64
Lysogeny, 123, 124
Lysokinase, 67, 69
Lysosoma, etc., 205, 206, 207, 208, 212
Lysozyme, 21, 111, 127, 205, 212

Mc
McDonald, 22
McFarlane, 58
McKay, 114
McMichael, 220

M

Mackay, 33
Maehder, 142
Maier, 183
Malmström, 72, 76

Malpighi, 47
Mamma carcinoma, 172, 177, 181, 185, 188, 197, 201, 208, 210, 214
Mann, 33
Marsden, 136, 146
Martin, 26, 28, 30, 33, 43, 44, 46, 62, 67
Martius, 160, 166
Marx, 163
Mason, 130
Matsumara, 81, 82
Maximow, 139
Megel, 28, 29, 33
Meggit, 136,
Meiers, 127, 134
Melanoma, 186, 196
Mellgren, 165
Menkin, 36, 46
Menten, 18 ó
Merkle, 158
Merten, 102, 103
Metastasis, 62, 64, 150, 169, 184, 210
Metschnikoff, 37
Michaelis, 17, 18, 165
Miller, 26, 75, 95, 100, 127, 134
Montagne, 117
Morawitz, 48
Moser, 78, 79, 84, 99
Moyniham, 130
Muld, 37, 46
Mullertz, 141
Murphy-Lymphoscarcoma, 156
Mutagene, 136
Mutation, 107
Mycosis fungoides, 196
Myrosin, 14

N

N-Acetyl-L-Tyrosine ethylester = (ATEE), 29
Nairin, 27
Nardi, 33
Nathanson, 220
Naylor-Foote, 31

Necrosin, 37
Neurath, 21
Niemann, 220
Nilsson, 63, 67
Nitrogen mustard, 161
Nomenclature, 16
Normal Substance, 137, 138
Northrop, 14, 25, 137
Novaline, 21
Nucleases, 16, 17, 120, 139, 141

O

Ogston, 33, 63, **68**
Ohler, 76
Old Age, 104
Ollendieck, 211
O'Meara, 62, 165, 166
Opie, 47, 52
Opsonin, 37
Ovarial carcinoma, 172, 174, 175, 178, 185, 187, 190
Ovarial sarcoma, 190
Oxidoreductase, 16, 17

P

PTC = (plasma thromboplastic component), 49
Pancreas carcinoma, 182, 186, 194
Papain, 21, 23, 31, 44, 141, 187
Papainase, 23, 39
Papayotin, 141
Parsons, 134
Pasteur, 14
Patel, 73, 76
Pearl, 114
Pepsin, 14, 16, 21, 23, 55, 60, 135, 141
Pepsinogen, 22
Peptidase, 51
Pereira, 134
Phagocytosis, 37, 40, 42
Phosphatase, 17, 206
Physick, 135, 146
Pistan, 33
Pisum sativum, 141

Pitt, 220
Plasma Ac-factor, 49
Plasma Ac-globulin, 49
Plasma-factor, X, 49
Plasma-Thromboplastic-Component (PTC), 49
Plasma-Thromboplastic-Antecedent, 50
Plasmathrombin-time, 75
Plasmakinin, 42, 49
Plasmin, 21, 28, 29, 31, 38, 39, 40, 43, 44, 47, 51, 52, 53, 54, 55, 56, 59, 60, 63, 64, 65, 67, 69, 71, 78, 79, 84, 106, 141, 157, 158, 159, 161, 205
Plasmin level, 40
Plasminogen, 31, 41, 47, 51, 52, 53, 55, 57, 60, 63, 65, 67, 69, 70, 71, 112, 205
Plasminogen activators, 41, 63, 80, 112
Platelet-co-factor, 49
Plazer-Altenburg, 177
Pliess, 118
Poliwoda, 72, 74, 76, 77
Polliack, 220
Popkin, 78, 79
Portio-Erosion, 191, 192
Preisig, 134
Proaccelerin, 49
Proactivator, 69
Proconvertin, 49
Pro-enzyme, 21, 47, 51
Prokipowicz, 80, 82
Prostate carcinoma, 63, 185, 192
Protaminsulphate, 155
Proteases, 16, 21, 23, 24, 31, 39, 40, 42, 44, 53, 54, 56, 64, 65, 67, 85, 88, 94, 95, 96, 98, 99, 108, 113, 117, 127, 128, 129, 131, 132, 139, 141, 145, 151, 156, 171, 187, 204
Proteinase, 16
Prothrombin, 40, 48, 49, 51, 52, 56, 62

Prothrombin time, 31
Prothrombinogen, 49
Prothrombokinase, 49
Pseudosubstrate, 19
PTA, 50, 66
Purden, 135
Purdom, 152, 164
Pusey, 136, 146
Pyrexin, 37
Q
Quick, 75

R
Raab, 81, 83
Rabbit, 45
Racematic substances, 19
Radicals (free), 114
Ransberger, 128, 134, 146, 160, 220
Rate of life, 114
Ravin, 22
Rawls, 130
Raynaud, 86
Reactivation, 152
Reaumur, 13
Rectum carcinoma, 181, 185, 193, 194, 202
Resorption of enzymes, 26
Respiratory Tract, 98
Reticuloendotheliosis, 178
Reticulo sarcomatosis cutis, 178
Ribonuclease, 28, 206
Rieser, 33
Ritter, 61, 67
Roberts, 151, 152, 164, 165
Rocha e Silva, 41
Roemheld, 102
Ronwin, 28, 33
Rosanova, 126, 134, 181
Rosenthal-factor, 50
Rossolek, 66, 68
Rosswick, 63, 67
Roth, 28
Ruhenstroth-Bauer, 127, 134, 152, 153, 164

Rupec, 220

S

SBI = (Soybean-trypsin-inhibitor), 29, 53
Saffiotti, 220
Sagiroglu, 142, 143, 146
Sailer, 65, 68, 73, 76
Sandritter, 66, 68
Sanz Anton, 183
Saphir, 151, 164
Sarcoma, 145, 182, 183, 186
Sautter, 81, 83
Sawyer, 62, 67, 72, 76
Schatten, 152, 164
Scheef, 116, 197, 205
Schmal, 127, 134, 150, 163
Schmidt, H. W., 65, 68
Schmidt, M. B., 150, 164
Schmidt, W. S., 153
Schmutzler, 72, 74, 76, 77
Schneider, 75, 76, 178
Schnellen, 178, 180
Schoneberger, 134
Schramm, 134
Schreck, 159, 165
Schulz, 75
Schumann, 113
Schwann, 14, 135
Schwert, 22
Schwick, 33
Seelig, 161, 165
Seligmann, 27, 28, 33
Serotonin, 41
Serum Ac-globulin, 49
Serum prothrombin, 49
Shaper, 62, 67
Shaw-McKenzie, 136, 146
Sheffer, 78, 79
Sherry, 29, 31, 33, 65, 68, 81, 83
Shields, 126, 134
Shulman, 55
Simpson, 164
Skin Ca., 196
Smyrniotis, 80, 81, 82
Smyth, 26, 27, 31, 33
Soybean-trypsin-inhibitor (SBI), 29, 53
Spallanzani, 13
Spar, 156, 165
Spleen, 141, 150
Spreading factor, 45, 152
Sproul, 162
Stabile factor, 49
Stasek, 91, 99
Stickiness, 127, 128, 154, 156, 162
Stieglitz, 117
Stojanow, 169, 173
Streptokinase, 31, 44, 47, 53, 64, 65, 66, 67, 69, 70, 71, 72, 73, 74, 75, 76, 79, 82, 84, 127, 205
Stuart-Prower-Factor, 50
Substitution therapy (in digestive disturbances), 24, 101, 102
Sumner, 14, 137
Sweet, 62, 67
Szillard, 106

T

Tagnon, 84, 99
Takahashi, 151, 164
Takeda, 58
TAME = (Tosyl-arginine-methylester), 29
Thiersch, 150, 164
Thies, 113, 118, 162, 165
Thomas, 117
Thornes, 159, 166
Thrombin, 29, 48, 51, 52, 53, 59, 60, 71
Thrombocytes, 40, 41, 52, 56, 61, 62, 64, 66, 150, 153
Thrombogen, 49
Thrombokatalysin, 49
Thrombokinase, 40, 48, 49, 51
Thrombolysis, 47, 61, 62, 63, 66, 67, 88, 170
Thrombophlebitis, 62, 88, 90, 91, 92
Thromboplastin, 49, 50, 75

Thromboplastinogen, 49
Thrombenzyme, 49
Thymus, 141, 187
Tillet, 47, 53, 69, 84, 99
Tilsner, 74
Tiralosi, 134
Titscher, 174, 200
Tobacco-Mosaic virus, 128
Todorutiu, 155
Tonsil carcinoma, 196
Tosyl-arginine-methylester = (TAME), 29
Transferases, 16, 17
Troll, 33
Trousseau, 162
Trypsin, 14, 16, 20, 21, 22, 26, 27, 28, 29, 31, 32, 44, 47, 53, 55, 60, 64, 65, 66, 67, 84, 85, 126, 127, 135, 136, 137, 141, 145, 155, 187, 205
Trypsin inhibitor, 29
Trypsinogen, 22, 53
Tryptophane, 55
Tsapogas, 64, 68
Turba, 22
Tween, 155
Tyrell, 134

U
Umbreit, 24
Ungar, 31
Uniform Theory, 36
Urease, 14, 137
Urogenital tract, 95
Urokinase, 29, 53, 64, 65, 67, 69, 70, 80, 81, 82, 84, 205
Uterus carcinoma, 172, 183

V
Vaginal carcinoma, 185
Valls-Serra, 85, 87, 88, 99
Van de Loo, 72
Varo, 92

Veermenko, 28
Verstraete, 71, 72, 76
Vitamin A, 131, 160, 200, 202, 205, 207, 208, 209, 210, 212, 213, 214, 215, 219
Vitamin-A-Intoxication, 207
Vitamin E, 187
Virchow, 35
Vocal cords, 150
Von Essen, 161, 165
Von Kaulla, 80, 82
V2-Carcinoma, 158, 159
Vulva carcinoma, 172, 183

W
Walford, 115
Walker Ca., 158
Walker-Carcinoma, 147, 158, 161
Walther, 151, 164
Warfarin, 158, 159, 160
Warts, 130
Webb, 24
Weidel, 126, 134
Weigelt, 142, 145
Weinbach, 112, 118
Werkmeister, 178
Werle, 31
Westrick, 180
Wild, 126, 134
Williams, 80, 82, 152, 164
Willis, 150, 151, 163
Winckelmann, 73, 76
Winkler, 182
Witte, 73, 77, 163
Wobe, 160, 211
Wobe-Mugos, 44, 45, 67, 85, 91, 93, 96, 97, 126, 128, 130, 158, 160, 168, 171, 172, 173, 174, 175, 177, 178, 179, 180, 182, 183, 187, 188, 189, 190, 191, 192, 193, 197, 198, 199, 200, 201, 202, 203, 204, 205, 210, 219
Wobenzyme, 44, 45, 85, 90, 93, 94, 130, 195

Wöhler, 14
Wohlman, 33
Wolf, 47, 54, 65, 88, 91, 99, 126, 128, 134, 138, 144, 158, 160, 167, 184, 220
Wolf-Zimper, 169, 171, 173
Wood, Jr., 151, 153, 154, 156, 157, 158, 159, 160, 161, 164
Wörn, 98
Woronski, 80, 82
Wrba, 174, 201, 211
Wyburn, 130

X
Xerion, 13

Y
Yoshida-Carcinoma, 160, 161
Yoshida-Sarcoma, 158

Z
Zeidmann, 150, 154, 163, 164
Zilber, 220
Zinser, 46
Zozeen, 13
Zymoplastin, 49